EXPLORING DISCRETE GEOMETRY

AMS / MAA | ANNELI LAX NEW MATHEMATICAL LIBRARY

VOL **56**

EXPLORING DISCRETE GEOMETRY

Thomas Q. Sibley

Providence, Rhode Island

EDITORIAL COMMITTEE

Bruce F. Torrence, Editor
Scott Cook
Courtney R. Gibbons
Beth Malmskog
Kenneth M. Monks
Japheth Wood

2020 *Mathematics Subject Classification.* Primary 52-01.

For additional information and updates on this book, visit
www.ams.org/bookpages/nml-56

Library of Congress Cataloging-in-Publication Data

Names: Sibley, Thomas Q., author.
Title: Exploring discrete geometry / Thomas Q. Sibley.
Description: Providence, Rhode Island : MAA Press, an imprint of the American Mathematical Society, [2024] | Series: Anneli Lax new mathematical library, 2578-6407 ; volume 56 | Includes bibliographical references and index.
Identifiers: LCCN 2024015227 | ISBN 9781470478070 (paperback) | ISBN 9781470478087 (ebook)
Subjects: LCSH: Discrete geometry–Textbooks. | AMS: Convex and discrete geometry – Instructional exposition (textbooks, tutorial papers, etc.).
Classification: LCC QA640.7 .S53 2024 | DDC 516/.11–dc23/eng/20240507
LC record available at https://lccn.loc.gov/2024015227

In memory of
Paul Fjelstad
creative spirit,
mentor, friend.

Contents

Introduction

> "When you have answered the
> question, it's time to question the
> answer."
>
> Paul Fjelstad

Good problems invite exploration and play, while needing perseverance. They also engender new problems and even new areas of mathematics. Discrete geometry developed in the twentieth century from problems that, while intriguing, seemed on the edge of traditional geometry. This book uses variations of a number of problems to lead to a deeper understanding of this relatively new area.

Discrete geometry studies arrangements of different numbers of points, lines, and other familiar objects, often looking for optimal arrangements or counting the number of ways of making these arrangements. Problems include counting distances determined by a set of points or placing "guard points" to "see" all other points in a given region. Others challenge us to divide a polygon into triangular regions or pack circles efficiently among many other problems. Many discrete geometry questions start from pure mathematical curiosity. More recently many of them have found application in computer imaging and other areas. Because of these applications, discrete geometry also involves algorithms to solve various aspects of these problems.

The answers to initial questions in this area often spark a variety of related questions. We fully embrace this idea of generating new questions as a unifying theme of this book, embodied in the Fjelstad quote above that starts our explorations. The following descriptions of the book's chapters illustrate this process.

- Chapter 1 provides a first layer of easily posed questions, and challenges you to play with them and perhaps even solve them.

- Chapter 2 answers the questions from Chapter 1, introducing some mathematical ideas along the way. Even more, it takes to heart the Fjelstad quote starting this introduction by posing new problems that are variations of the problems of Chapter 1.

- Chapter 3 answers the problems of Chapter 2, provides related mathematical ideas and, continuing the theme of the book, questions those answers with more variations.

- Chapter 4 brings closure by answering the variations of Chapter 3 and indicating a broader view, including what is known about some of the topics in this area.

- The material after Chapter 4 starts with answers to the exercises not answered earlier. After that are suggestions of books and articles at an accessible level for further exploration, together with a short description of each one. (Of course, you can find much more information on these topics on many sites on the internet. However, determining whether a given site is a good source is much harder.) We use the mathematical form of referencing in the text, such as [6, 21], which refers to page 21 of the sixth item (the book *Excursions into Mathematics*) in the Bibliography section, which appears after the Suggested Readings. The Index rounds out the end material.

This book's problems seek to challenge anyone excited about mathematics and doesn't assume any background beyond high school mathematics. The explanations also require no more background, although the topics range far beyond the high school curriculum. We assume a familiarity with common geometrical and algebraic concepts, rather than burdening the text with a more formal presentation. We will define less-common terms, which appear in italics in their definition. The reasoning will vary from computations and informal arguments to more careful, but still accessible, proofs to fit each context.

Discrete geometry can improve valuable visualization skills because it asks different kinds of questions from those asked in traditional geometry classes. The problems in discrete geometry, like any area of mathematics, will also improve your problem solving skills, a valuable asset for anybody. The process of "questioning the answer" and the variations it spawns raises problem solving to a new level. From a mathematician's point of view, the variations provide more important benefits: they help deepen mathematical reasoning and geometrical intuition, and they introduce one way mathematicians find new research questions. Indeed, some variations of problems we consider lead to open research questions or recent results. Mathematical research often develops from seeing patterns and making conjectures about why those patterns happen. I hope you will grow mathematically by wrestling with the variations in this book, seeing patterns in them, making conjectures about them, solving them, and even making up your own variations to explore! But the main motivation for writing this book is that I found these problems and variations fun to play with—and I hope you find much pleasure in pondering them as well.

Acknowledgments

My thanks to the Anneli Lax New Mathematical Library Editorial Board of the Mathematical Association of America and especially the chair, Bruce Torrence, for many helpful suggestions and guidance. I greatly appreciate the expertise and professionalism of my editor, Sergei Gelfand, and his team from the American Mathematical Society. My wife, Jennifer Galovich, has given the loving support and sympathetic ear that I needed and valued throughout the writing. I remember with deep fondness the inspiration of my college mentor, Paul Fjelstad, whose creativity and enthusiasm guided me throughout my career. This book is dedicated in his memory.

1

Beginning Explorations

Discrete geometry, like all of mathematics, has developed piecemeal. It is still new enough to have unsolved problems that seem just a few steps from introductory ones. The problems of this chapter seek to whet your appetite. Think about them—draw pictures for small cases, look for patterns, make conjectures, and try to solve them. Complete answers are great, but so are partial answers, guesses, and even mistakes. Recall: mathematics is not a spectator sport; so enjoy engaging with these problems. We'll revisit these problems in later chapters, providing answers, relevant mathematics, and especially new questions probing these areas more deeply.

1.1 Lines and Regions

A line drawn on a sheet of paper splits the paper into two regions. Two lines determine either three or four regions, as indicated in Figure 1.1.

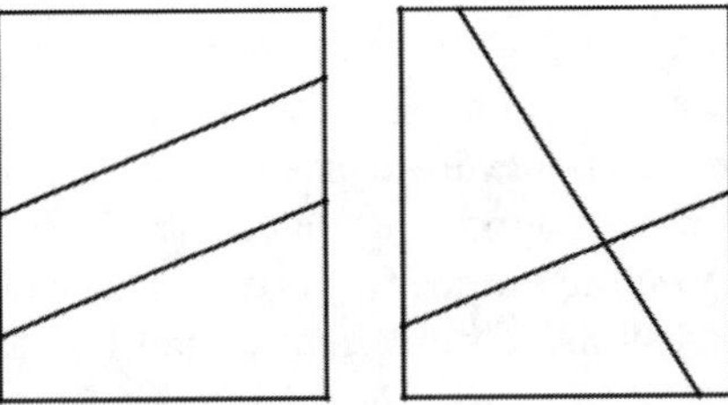

Figure 1.1. Two lines determine three or four regions.

Problem 1.1. What is the largest number of regions in a plane that three lines, four lines, or in general n lines determine? What is the smallest number of regions that n lines determine?

1.2 Diagonals and Triangulations

A triangle has no diagonals, a square has two and a convex pentagon has five, as in Figure 1.2. By "convex" we mean that it doesn't have any dents—see Figure 1.3. We define convex and diagonal after Figures 1.2 and 1.3. We often abbreviate "a polygon with n sides" as an *n-gon*. We'll call the corners of a polygon *vertices* and the sides *edges*, names that mathematicians use in broader contexts. (Your intuition of what a polygon is suffices for this book. However, the Appendix after Chapter 1 explores the thinking leading to a more formal definition.)

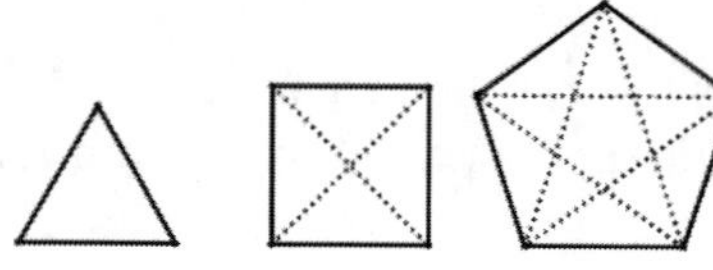

Figure 1.2. Diagonals of a triangle, a square, and a convex pentagon.

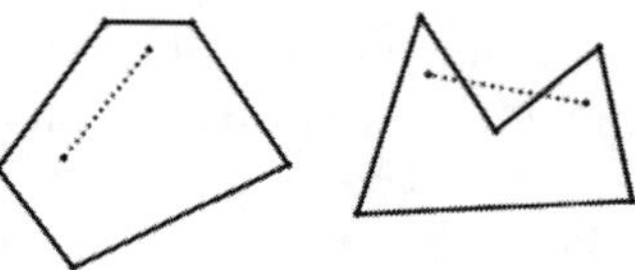

Figure 1.3. A convex shape and a nonconvex shape.

Definition. A set of points is *convex* provided for any two points P and Q in the set, the line segment $\overline{PQ}$ is entirely in the set. Otherwise, the set is *nonconvex*. We say a polygon is convex whenever its interior is a convex set.

Definition. For two vertices P and Q of a polygon, the line segment $\overline{PQ}$ is a *diagonal* provided it is not an edge of the polygon and, except for the endpoints P and Q, $\overline{PQ}$ is entirely in the interior of the polygon.

If a polygon is convex, whenever we pick two vertices not next to one another (not *adjacent*), the segment connecting them is a diagonal. So the number of diagonals depends only on the number of vertices, suggesting Problem 1.2. Once we ask about convex polygons, it is only natural to think about nonconvex ones, the topic of Problem 1.3.

Problem 1.2. How many diagonals does a convex polygon with n vertices have?

Problem 1.3. Do nonconvex polygons with at least four vertices always have some diagonals? If so, do all n-gons have a minimum number of diagonals (that may depend on n)?

From Figures 1.2 and 1.3, we can split the square into two triangles and each pentagon into three triangles, as in Figure 1.4.

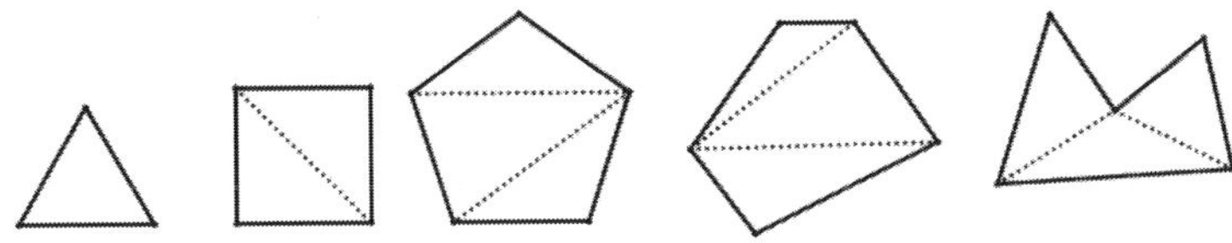

Figure 1.4. Triangulations of polygons from Figures 1.2 and 1.3.

Problem 1.4. Can we "triangulate" any polygon, as defined below, whether or not it is convex? Does every triangulation of an n-gon have the same number of triangles?

Definition. A *triangulation* of a polygon is a collection of triangles with nonoverlapping interiors, whose vertices are vertices of the polygon so that the triangles and their interiors together cover all of the polygon and its interior.

1.3 Distances and Points

Problem 1.5 (a). What is the smallest number of different (nonzero) distances n points on a line determine? What is the largest number? Can n points on a line determine any number of different distances between the smallest and largest numbers?

Problem 1.5 (b). Is there an arrangement of some number n of points in the plane that determine fewer distances than the smallest number of distances of n points in a line? What is the smallest number of distances 3, 4, 5, 6, 7, or 8 points in the plane can determine?

1.4 The Art Gallery Problem

In a museum or art gallery, guards (or nowadays cameras) are posted so that together they can see all the artwork to thwart thieves. If the museum or gallery were convex, one guard could survey all the art at once (provided there weren't people in the way). But real galleries and museums have far more complicated shapes. What is the fewest number of guards needed for any shaped polygons, based on the number of vertices? Figure 1.5 gives two examples. We state this more formally in Problem 1.6. The problem assumes the guards (the points G_i) can see in all directions, but not through the edges (walls) of the polygon.

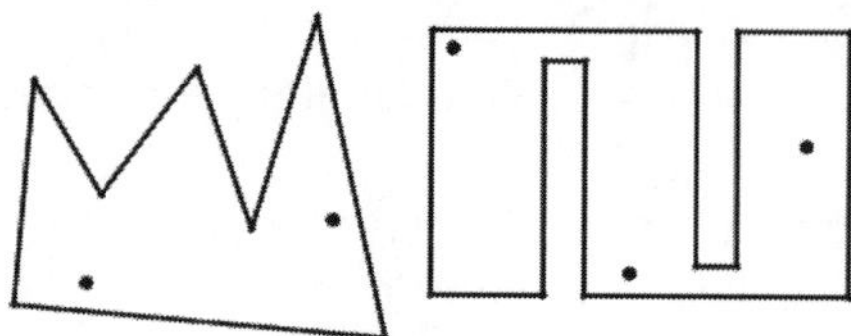

Figure 1.5. Art galleries needing two or three guards.

Definition. The points $G_1, G_2, \ldots, G_k$ form a set of *guard points* for a region R provided that for all points P of R, there is some guard point G_i with the segment $\overline{G_i P}$ entirely contained in R.

Definition. The minimum number of guard points needed for any n-gon is g_n.

Problem 1.6. Explain why $g_4 = 1$, even though there are nonconvex quadrilaterals. Find g_5. Find a polygon with the fewest number of vertices needing two guard points. Repeat for three or more needed guard points. Look for a pattern for g_n.

1.5 Geometric Patterns

People throughout history and all over the world have delighted in designs, including colored patterns. Mathematicians have worked on classifying the possible types of patterns, colored or not. Figure 1.6 gives three two-colored patterns and a three-colored pattern for a regular hexagon that exhibit some sort of regularity in its coloring pattern, whereas the coloring of regions in Figure 1.7 seems irregular. (That is, it might be difficult to describe the pattern of how the colors are arranged in Figure 1.7.) In addition to looking for more such patterns in Problem 1.7 (a), Problem 1.7 (b) challenges you in a different way: exploring what we mean by regularity in coloring. The abstract nature of mathematics allows us to define concepts any way we want, but arbitrary definitions rarely prove worthwhile. Good definitions capture some key aspects of what we are studying and lead to interesting results.

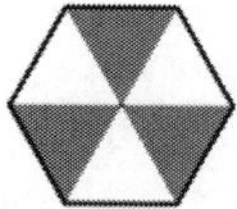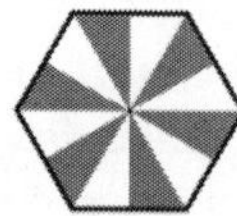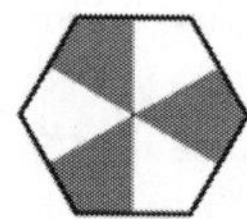

Figure 1.6. Regular colorings of a hexagon.

Figure 1.7. An irregular coloring of a hexagon.

Problem 1.7 (a). Design colored patterns for regular polygons with different numbers of vertices exhibiting regularity with two or more colors. What possible numbers of colors can such a coloring have in an n-gon?

Problem 1.7 (b). Describe what you mean by a pattern having a regular coloring.

1.6 Voronoi Diagrams

Given some fixed points in the plane, called *sites*, many people have investigated the regions of points closest to each of the fixed sites, starting at least with René Descartes (1596–1650). The resulting pictures, as in Figure 1.8, are called Voronoi diagrams, after the Ukrainian mathematician Georgy Feodoseevich Voronoi (1868–1908) who generalized them shortly before his death. Since then, scientists in disciplines ranging from astronomy to biology to engineering have found numerous applications of them.

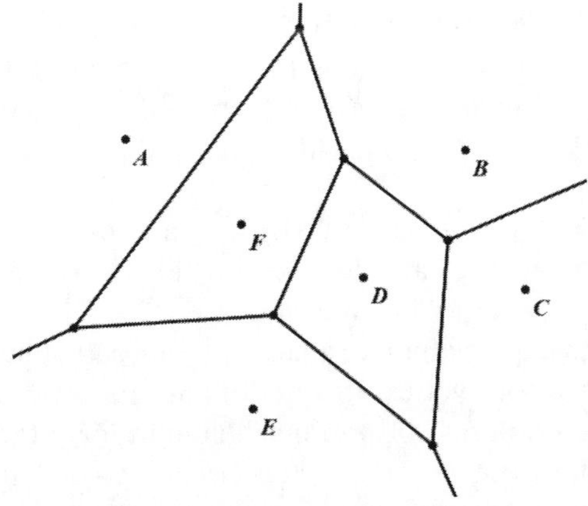

Figure 1.8. The Voronoi diagram for the six labelled points.

Definition. Given sites S_1, S_2, ..., S_n in the plane the *Voronoi region* of S_i is the set of all points X so that for all sites S_k with $k \neq i$, $d(S_i, X) < d(S_k, X)$, where $d(X, Y)$ is the (usual Euclidean) distance between the points X and Y. The Voronoi regions together with their boundaries form the *Voronoi diagram* of the sites.

Problem 1.8. Find Voronoi diagrams when there are two, three, and four sites. What property or properties in terms of the given sites characterize the line segments, lines, and rays dividing the regions of a Voronoi diagram?

Remark. Along with puzzling about the preceding problems, you might be wondering "What about these geometric topics make them discrete?" "Discrete" means "separate," and in mathematics it is the opposite of continuous. For example, the lines in Problem 1 create separate regions, things we can count. Some of the other problems also asked you to count geometric items. In Problems 7 and 8 you didn't count anything, but you did color or construct discrete regions. Lines, planes, and many other geometric objects have continuous aspects, but these problems and almost all of those in Chapters 2, 3, and 4 focus on their discrete properties.

Appendix. What Is a Polygon?

Elementary school students develop the idea of a polygon from seeing lots of pictures of examples—and usually convex ones at that. Such an intuitive understanding of a polygon suffices for this book, but a formal definition is more involved. Mathematicians require careful definitions to prove results. They really don't like the phrase "the exception proves the rule" since they never want to allow a theorem to have any exceptions. So they look for potential *counterexamples*: things that are close to the intuitive idea of what they are trying to define, but fail in some way. The definition they craft needs to exclude the counterexamples while still including everything that fits their intuition. Let's use Figure 1.9 to compare the example of a polygon there with several designs that violate the common intuition of a polygon. After that we'll formulate a mathematical definition of a polygon.

A polygon is made up of vertices (corners) and edges (sides). Further, we expect a polygon to have a clear interior. The polygon on the left of Figure 1.9 has five vertices, five edges, and an interior.

The second drawing has four vertices and four edges, but two of the edges cross at a point that is not a vertex. Its interior appears as two triangular regions. A satisfactory definition of a polygon needs to eliminate examples with edges that intersect in a point other than a vertex. (In fact, we will use that property in the proof of Theorem 2.1 in the next chapter.) The third illustration adds a vertex at the intersection of the two previous edges, which seems to get around the

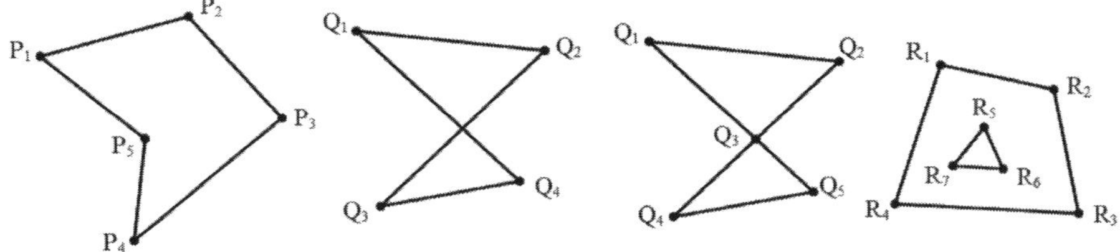

Figure 1.9. A polygon and three nonpolygons.

problem of crossing edges. But it is really two triangles sharing a vertex, rather than one polygon. So, we need to eliminate that option as well. In the design at the right, no edges cross, but it seems better described as a polygon with another polygon inside it, rather than just one polygon. (In Chapters 3 and 4, we'll consider polygons with holes like this.) Before we define a polygon, we define a segment of a line so that we can define an edge of a polygon.

Definition. The *segment* $\overline{PQ}$ is the set of points on the line through P and Q that are between P and Q as well as P and Q, which are called the *endpoints* of the segment.

Definition. An n-sided *polygon* (n-gon) is a set of n distinct points (called *vertices*) P_1, P_2, ..., P_n in the plane and the n segments (called *edges*) $\overline{P_1 P_2}$, $\overline{P_2 P_3}$, ..., $\overline{P_{n-1} P_n}$, $\overline{P_n P_1}$ so that the intersection of two edges is either empty or is their common endpoint.

Check that the three counterexamples of Figure 1.9 fail to satisfy our definition of a polygon, but the pentagon of Figure 1.9 does satisfy it. Draw some figures you think should be polygons and some you think should not be polygons. Test which fit the definition above.

While it may seem obvious that polygons have an interior, mathematicians found a proof of this quite challenging. Camille Jordan (1838–1922) gave the first proof of the general theorem in 1887, which is now called the Jordan curve theorem in his honor. He showed that polygons and more generally "simple closed curves" separate the plane into an exterior and an interior. The theorem also shows how to tell whether two points are both inside or both outside the polygon (or curve). It does so by drawing the line segment between them and counting how many times it crosses the polygon. If it crosses an even number of times, they are both inside or both outside; if an odd number of times, one is inside and one is outside. It helps to have one of the points clearly outside or clearly inside. In Figure 1.10, the point C is clearly outside the polygon. Since the line segment $\overline{BC}$ crosses four edges of the polygon, point B is also outside. The segment $\overline{AC}$

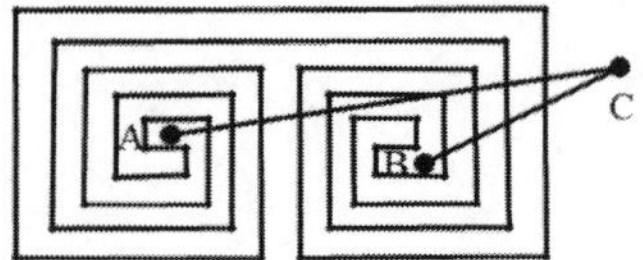

Figure 1.10. A polygon with some points inside and outside.

crosses nine edges, so A is inside the polygon. (See [7, 244–246] for more on this theorem.)

In addition to polygons, geometers have explored polygrams, also called star polygons. The best known is a pentagram, shown on the left of Figure 1.11. In polygrams, the edges cross in a set way, but otherwise they satisfy the definition of a polygon. (The numbering of the vertices of the pentagram enable the edges to fit the definition of a polygon, ignoring the edge crossing.) The second design of Figure 1.11 is not a polygram since it is made from two cycles of edges. No labeling of the vertices will make the edges fit the definition of a polygon, even ignoring the crossings. The second design is called a compound polygon. The last two drawings in Figure 1.11 are two different kinds of heptagrams. Note that the vertices of the two heptagrams are numbered differently so as to fit the labeling of the edges in the definition of a polygon. ("Hepta" is Greek for seven.)

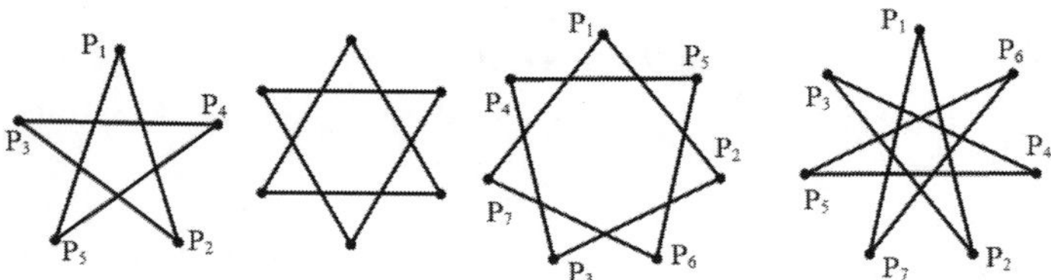

Figure 1.11. A pentagram, a compound polygon, and two heptagrams.

Exercise 1.1. Draw other polygrams and compound polygons whose vertices are the vertices of a regular n-gon. Look for conditions determining whether you get a polygram or a compound polygon. What conditions do we need to set on the crossings so that the second drawing in Figure 1.9 is not a polygram?

The answer to this exercise and the exercises in later chapters come after Chapter 4.

2

First Variations

Curiosity together with engaging problems propels mathematics forward. The following answers to the Chapter 1 problems suggest further problems to investigate. Each section follows the pattern of restating the related Chapter 1 problem or problems, answering them, and then questioning those ideas in at least one way. I encourage you also to try your hand at asking and answering your own variations of the problems.

2.1 Lines and Regions

Problem 1.1. (repeated) What is the largest number of regions in a plane that three lines, four lines, or in general n lines determine? What is the smallest number of regions that n lines determine?

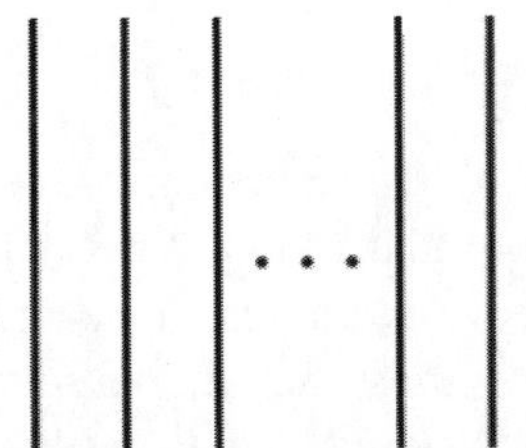

Figure 2.1. n parallel lines determine $n + 1$ regions

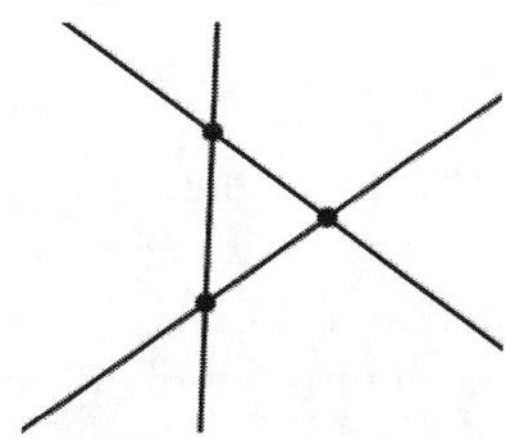

Figure 2.2. Three general lines determine seven regions.

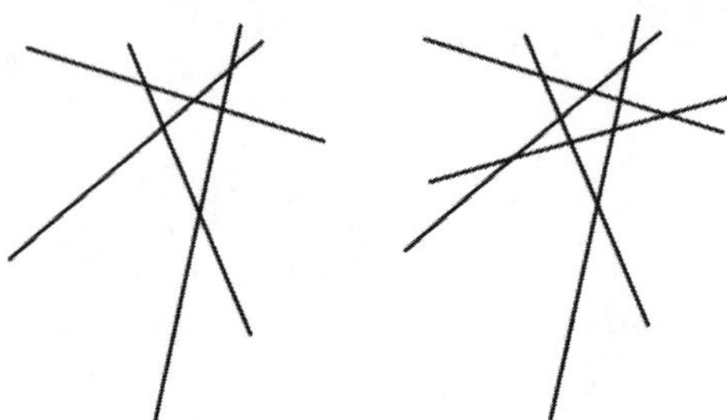

Figure 2.3. Regions determined by four and five lines in general position.

A set of n lines in the plane determines the fewest regions when they are all parallel to each other. They determine $n+1$ regions: $n-1$ slices between adjacent lines and the two half-planes on the ends, illustrated in Figure 2.1. In Figure 2.2, we see that three lines can determine seven regions. We say these lines are in *general position*—as opposed to the special cases of some parallel lines or three or more lines intersecting in the same point.

How can we find the maximum number of regions that n lines can determine? Draw some pictures to convince yourself first that to get the maximum number of regions, we need the lines to be in general position. Next, let's look at the values for $n = 1, 2, 3, 4$, and 5 lines to see whether we can find a pattern. Figure 2.3 gives examples of general cases for $n = 4$ and $n = 5$. Table 2.1 lists the maximum number of regions.

Table 2.1. Maximum number of regions for $n \leq 5$ lines.

lines	1	2	3	4	5
regions	2	4	7	11	16

The values in Table 2.1 suggest a pattern: the nth line adds n more regions to what the maximum was before. This describes the maximum number of regions *recursively*, that is, values are determined using previous values. If we let $M(n)$ be the maximum number of regions determined by n lines, the pattern seems to be $M(n) = M(n-1)+n$. We will consider looking for a *closed form formula* for $M(n)$ shortly; that is, a formula depending only on n, the number of lines, not needing a previous value like $M(n-1)$. But first there is a more important mathematical question: why is the recursive pattern correct—why does the nth line add at most n regions?

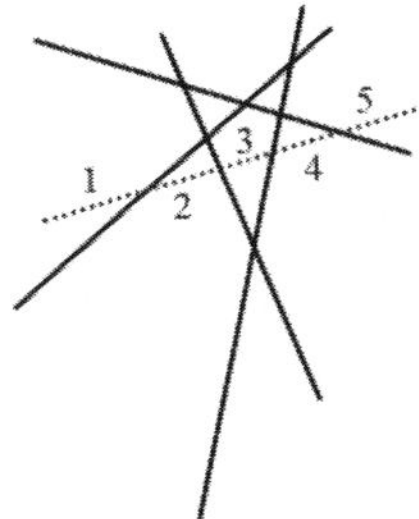

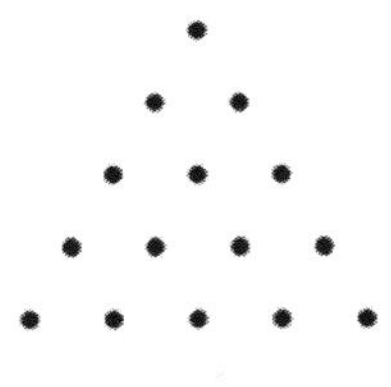

Figure 2.4. The pieces on the added line.

Figure 2.5. Triangular arrangement of points.

Suppose we have $n - 1$ lines making the maximum number of regions. In general position, the nth line intersects each of the other $n - 1$ lines once. So, the new line is divided into n parts, including the two infinite ones, as labeled in Figure 2.4, where the dotted line is the new one. Note that each of the pieces of the new line divides one previous region in two, giving n new regions, as predicted. This argument also shows the importance of having the lines in general position. If the new line is parallel to any of the previous lines, we will miss an intersection and so fail to divide a region. Also, if the new line goes through the intersection of two or more previous lines, we will have fewer new regions. Thus $M(n)$ equals $M(n - 1) + n$.

The numbers of regions in Table 2.1 are one more than the better known sequence of *triangular numbers* $1, 3, 6, 10, 15$. These numbers come from stacking points or balls in a triangular array. The nth triangular number counts the number of points in the top n rows of Figure 2.5. The term "triangular numbers" and the even better known term "square numbers" have been around since the ancient Greeks, who also knew a formula for triangular numbers. (However, they had to describe the formula in words since our familiar algebraic notation is much more recent.) Figure 2.6 helps visualize the formula: the nth triangular number is half of a "*rectangular number*," where one side of the rectangle is one longer than the other side. The formula for a rectangular number is $n(n + 1)$. In turn, the nth triangular number is half of that, $\frac{n(n+1)}{2}$. Hence, the maximum number of regions in the plane determined by n lines is $M(n) = \frac{n(n+1)}{2} + 1 = \frac{n^2+n+2}{2}$.

One of the pleasures of mathematics comes after we answer a question—it is now "time to question the answer." That is, we modify the question to lead us in new directions and perhaps gain deeper insight. Here are several variations on Problem 1.1 for you to explore. You can invent your own variations of these problems and all of the others and try to answer them as well.

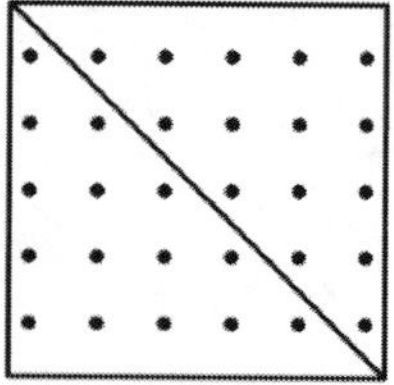

Figure 2.6. A rectangular number as the sum of two triangular numbers.

Problem 2.1 (a). By using some placement besides general position or all parallel lines can we have any number of regions between the minimum and maximum number of regions in the plane determined by n lines?

Problem 2.1 (b). What are the maximum and minimum number of regions in three-dimensional space determined by n planes?

Problem 2.1 (c). What are the maximum and minimum number of regions in the plane determined by n circles?

Problem 2.1 (d). What are the maximum number of regions in the plane determined by two convex polygons with j and k sides?

2.2 Diagonals and Triangulations

Problem 1.2. (repeated) How many diagonals does a convex polygon with n vertices have?

We can start as we did for Problem 1.1 about regions determined by lines—Table 2.2 lists the number of diagonals for different-sized convex polygons—and look for patterns. This problem illustrates one of the intriguing and beautiful aspects of mathematics: we can often solve a problem in different ways that yield the same answer.

Table 2.2. Number of diagonals in a convex polygon with n sides.

sides	3	4	5	6	7
diagonals	0	2	5	9	14

The difference in the number of diagonals goes up by one each time: $2 - 0 = 2$, $5 - 2 = 3$, $9 - 5 = 4$, etc. These differences showed up with Problem 1.1. Indeed, the values in Table 2.2 are one less than triangular numbers, although the n here doesn't match with the number k of numbers being added. Table 2.3 combines Table 2.2 with the corresponding triangular numbers.

Table 2.3. Relation of diagonals and triangular numbers.

sides (n)	3	4	5	6	7
diagonals	0	2	5	9	14
k	1	2	3	4	5
sum of first k numbers	1	3	6	10	15

In Table 2.3, $k = n - 2$. From the formula for triangular numbers, $k(k+1)/2$, we derive the number of diagonals: $\frac{(n-2)(n-1)}{2} - 1 = \frac{n^2 - 3n + 2 - 2}{2} = \frac{n(n-3)}{2}$.

Why is this formula correct? There are n vertices and each one is connected by a diagonal to each vertex that isn't adjacent to it. There are $n - 3$ of those: we subtract the vertex and its two neighbors from all n of the vertices. That would give us $n(n-3)$ diagonals. But we have counted each diagonal twice, once starting from each of its two ends. That explains why we need to divide by 2 to get the correct formula.

You might also note that if we added the columns of Table 2.2 as we do in Table 2.4 the entries in the row of total edges are triangular numbers but shifted by one. That is, a convex polygon with n sides has $\frac{n(n-1)}{2}$ total edges. The sides together with the diagonals give all possible segments or edges between pairs of vertices of the polygon. Counting all the ways of choosing (unordered) pairs from n things occurs in many contexts. Mathematicians developed a notation for this, $\binom{n}{2}$, read "n choose 2" or more formally the combinations of n things 2 at a time. (Over a thousand years ago, mathematicians generalized the concept to $\binom{n}{k}$, the combinations of n things k at a time and found ways to compute these values. Calculators use nCk for this value. A general formula needed modern notation and so is more recent—a few hundred years old. We don't need the general formula, but you can find it online or in [28, 239–242].)

Table 2.4. Sides, diagonals, and total edges of a convex polygon.

sides n	3	4	5	6	7
diagonals	0	2	5	9	14
total edges $\binom{n}{2}$	3	6	10	15	21

The total number of sides and diagonals of a convex n-gon is the sum of the first $n - 1$ numbers. And that value is n bigger than the number of diagonals because there are n sides to the polygon. The algebra gives us the same formula for the number of diagonals: $\frac{(n-1)n}{2} - n = \frac{n^2 - n - 2n}{2} = \frac{n(n-3)}{2}$.

Table 2.5. Ratio of diagonals to sides.

sides	3	4	5	6	7
diagonals	0	2	5	9	14
$\frac{\text{diagonals}}{\text{sides}}$	0	$\frac{1}{2}$	1	$\frac{3}{2}$	2

Table 2.5 gives yet another pattern from Table 2.2.

The ratio of diagonals to sides goes up by $\frac{1}{2}$, giving $\frac{n-3}{2}$. If we multiply by n, the number of sides, we again get $\frac{n(n-3)}{2}$, the number of diagonals in a convex n-gon. Beautiful: three ways to approach counting the diagonals, all leading to the same answer.

The three-dimensional analog of a polygon is a *polyhedron*, Greek for "many faces." Examples of polyhedra (the plural of polyhedron) include a cube, pyramids, prisms, and many more shapes. The surface of a polyhedron is made of polygons, its *faces*. Every edge of one of these polygons is shared with just one other polygon, and at least three faces meet at each vertex so that the shape is three-dimensional. A formal definition of a polyhedron goes beyond the level of this book. See [11, 162] for one way to define a polyhedron. (Other researchers use somewhat different definitions, depending on what aspect they are studying.)

Diagonals for polyhedra present more challenges than diagonals for polygons. First of all, Figure 2.7 illustrates the difference between diagonals on the surface of a polyhedron (a cube here), such as $\overline{AB}$, and interior diagonals, such as $\overline{AC}$. Each face of a polyhedron is a polygon, and so there will be surface diagonals unless all the faces are triangles. Pyramids, as in Figure 2.8, have no interior diagonals.

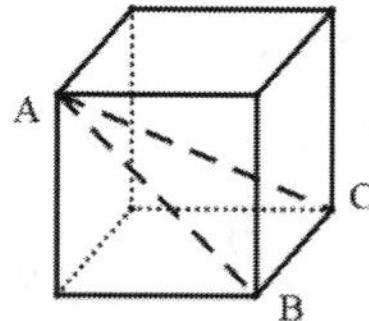
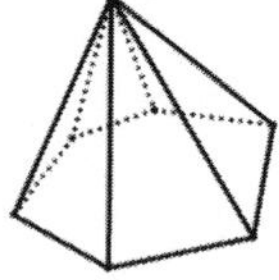

Figure 2.7. Surface and interior diagonals.

Figure 2.8. A pyramid has no interior diagonals.

Problem 2.2. Does every convex polyhedron that isn't a pyramid have an interior diagonal?

Problem 1.3. (repeated) Do nonconvex polygons with at least four vertices always have some diagonals? If so, do all n-gons have a minimum number of diagonals (that may depend on n)?

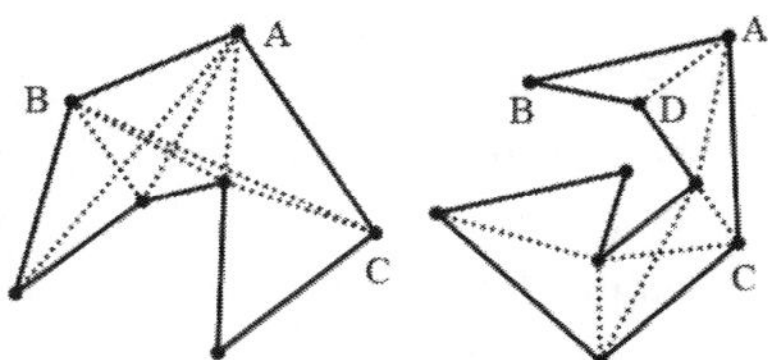

Figure 2.9. Diagonals of nonconvex polygons.

After you draw some nonconvex polygons, as in Figure 2.9, you will probably convince yourself that they must all have some diagonals, although the minimum number might not be obvious.

Theorem 2.1. Every polygon with at least four vertices has at least one diagonal.

Proof. Pick a direction and find the vertex, call it A, furthest in that direction. (If there are two or more tied for furthest, modify the direction.) In Figure 2.9, we pick the highest point in each polygon. The point A has two edges from it to the vertices we'll call B and C. Since there are at least four vertices, $\overline{BC}$ is not an edge of the polygon. There are two possibilities. As in the left polygon, $\overline{BC}$ is a diagonal and we are done. Otherwise, as in the right polygon, inside triangle $\triangle ABC$ there is some vertex D that is closest to A. Claim: $\overline{AD}$ must be a diagonal. To see this, think of what could go wrong. There would be some edge $\overline{EF}$ of the polygon crossing the segment $\overline{AD}$. The edge $\overline{EF}$ can't cross either of the edges $\overline{AB}$ and $\overline{AC}$ because a polygon's edges never cross one another. (See the Appendix after Chapter 1 for properties of a polygon.) As a consequence, at least one of E and F would be inside the triangle $\triangle ABC$. As Figure 2.10 illustrates, one of E or F would have to be closer to A than D is, a contradiction. Thus $\overline{AD}$ is in the interior of the polygon. Since only B and C are adjacent to A, $\overline{AD}$ must be a diagonal. $\qquad\square$

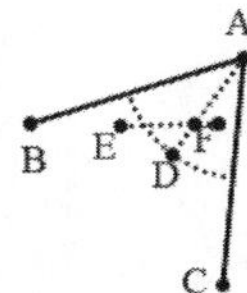

Figure 2.10. An impossible scenario provided D is the closest vertex to A.

Remarks. We ended this proof with the end of proof symbol □. In older times, proofs often ended with the initials Q. E. D., an abbreviation for the Latin phrase "quod erat demonstrandum," meaning "which was to be proven (demonstrated)." This is the first result we have called a theorem and for which we gave an explicit proof. Earlier results came from computations or were a fairly direct outcome of understanding the terms and figures and so didn't require careful examination. We'll reserve theorems and proofs for statements that aren't so easily grasped or which require more careful arguments. The latter part of this argument (after the phrase "think of what could go wrong") is an example of a proof by contradiction. For more on this ancient technique, see [28, 75–76]. (Aristotle (384–322 BCE) gave a proof using this technique over 2300 years ago.)

Now that we know a polygon with at least four vertices has a diagonal, we focus on the minimum number of diagonals. Figure 2.11 suggests that an n-gon has at least $n - 3$ diagonals. The proof of this in Theorem 2.2 uses induction, a powerful technique to prove statements about positive integers. Induction starts by showing a first value holds, called the base case. (For us the base case will be a quadrilateral, the smallest shape that has diagonals.) Then we show how to prove an additional step from the step where we are using previously proved steps. (We use quadrilaterals to show pentagons work and use quadrilaterals and/or pentagons to show hexagons work, etc.)

Induction proofs depend logically on what we mean by the positive integers or counting numbers. When we say "1, 2, 3, and so on," our intention is that each counting number n has a unique successor $n + 1$, starting with one, the first positive integer, and the process of adding one each time will get us to all the positive integers. That is, once we get started and know how to keep taking another step, we can "go all the way," so to speak. (See, for instance, [28, 93–102] for an in depth discussion of induction.) While induction isn't as venerable as proofs by contradiction, the idea of an induction proof is at least 700 years old.

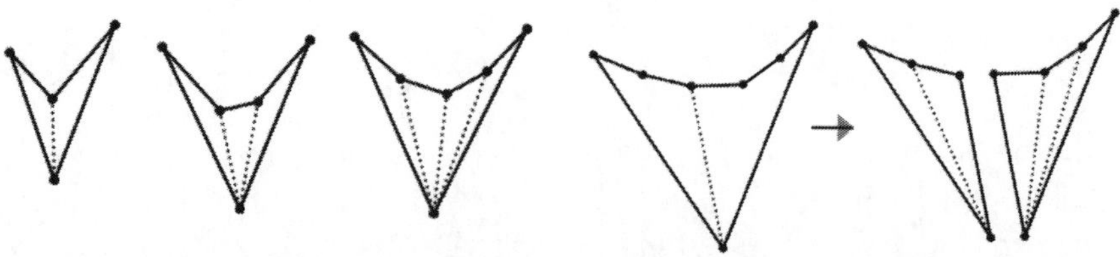

Figure 2.11. Nonconvex polygons with a minimum of diagonals.

Figure 2.12. Splitting a polygon into two polygons.

Figure 2.12 illustrates the key idea of the proof of Theorem 2.2: to determine the minimum number of diagonals for a larger polygon, we use one of its diagonals to split it into two smaller polygons and count their diagonals. In Figure 2.12 the smaller polygons have one and two diagonals. Thus the bigger polygon has four diagonals: those three plus the dividing diagonal of the bigger polygon. Mathematicians often use this idea of reducing a problem to an easier, already solved problem beyond its use in induction proofs.

Theorem 2.2. Every polygon with n vertices, for $n \geq 4$, has at least $n - 3$ diagonals.

Proof by induction. The statement is true by Theorem 2.1 for the initial case of $n = 4$. Suppose for $k \geq 4$ that every polygon with at most k vertices satisfies our claim. We'll extend this to polygons with $n = k + 1$ vertices. From Theorem 2.1 a polygon with $k + 1$ vertices has at least one diagonal $\overline{XY}$. Use this diagonal to separate the polygon into two polygons each with fewer vertices, say with i and j vertices, where $3 \leq i \leq k$ and $3 \leq j \leq k$. By our supposition, these have at least $i - 3$ and $j - 3$ diagonals, respectively. We now have $1 + (i - 3) + (j - 3) = i + j - 5$ diagonals. Both of these smaller polygons have X and Y as vertices, so they are counted twice: $i + j = n + 2$. But then we have at least $i + j - 5 = n - 3$ diagonals, as desired. By induction, the claim holds for every $n \geq 4$. $\qquad\square$

What else can we ask about diagonals? How about the range of diagonals a polygon can have?

Problem 2.3. For every number k between $n - 3$ and $\frac{n(n-3)}{2}$, is there an n-gon with k diagonals?

The ideas of induction and using a diagonal to split a polygon into two smaller polygons, as in the solution to Problem 1.3, give us a way to solve Problem 1.4 as well.

Problem 1.4. (repeated) Can we "triangulate" any polygon, whether or not it is convex? Does every triangulation of an n-gon have the same number of triangles?

Theorem 2.3. Every n-gon has a triangulation with $n - 2$ triangles and all triangulations of an n-gon have $n - 2$ triangles.

Proof by induction. Every polygon has at least three vertices, so the base case is about triangles. But a triangle is already a triangulation with $n = 3$ and $n - 2 = 1$ triangle. With only three vertices, there is only one way to triangulate it. For $k \geq 3$, suppose that every polygon with j vertices and $3 \leq j \leq k$ has a triangulation with $j - 2$ triangles and that every triangulation has $j - 2$ triangles. Given a polygon with $n = k + 1$ vertices, use a diagonal to split it into two polygons sharing the diagonal with i and j vertices. As before, $i + j = n + 2 = k + 3$ vertices and both i and j are between 3 and k. Any triangulation of the one with i vertices has $i - 2$ triangles and the other has $j - 2$ triangles in its triangulation. Together these give a triangulation of the big polygon with $i + j - 4 = k - 1 = (k + 1) - 2$ triangles, just the number we need. Further, we can construct every triangulation of a $(k + 1)$-gon following the procedure we used for counting triangles. By induction, the claim holds for all polygons.   □

Consider another way of looking at the difference in the number of vertices of a polygon and the number of triangles in a triangulation of it. Not every triangle in the triangulation can have just one edge of the polygon since there are two more edges than triangles. In fact, if the polygon has at least four vertices, at least two of the triangles have to have two edges of the polygon with their third side a diagonal. Corollary 2.4 summarizes this situation. Figure 2.13 illustrates why these triangles are fancifully called ears. The polygon in Figure 2.13 has more than two ears, as defined below, but the shaded ones look like ears. (Mathematicians use the name "corollary" for a theorem that follows easily from an earlier result.)

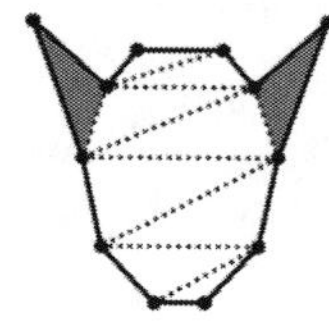

Figure 2.13. A polygon with two ears shaded in.

Definition.   In a polygon with vertices V_1, V_2, ..., V_n, a triangle $\triangle V_{i-1}V_iV_{i+1}$ is an *ear* provided that the polygon has a diagonal from V_{i-1} to V_{i+1}.

Corollary 2.4 (Two ears theorem). Every polygon with at least four vertices has at least two ears.

Proof. Triangulate the polygon. There are two fewer triangles than edges. No triangle can have all three of its sides as sides of the polygon since the polygon is not a triangle. So two (or more) of the triangles in the triangulation have to have two edges and a diagonal. These are ears.   □

What happens to the number of triangles if there are interior points that need to be included as vertices in a triangulation? Figure 2.14 gives two triangulations of a hexagon with two interior points, and these triangulations have the same number of triangles, eight. Note that by Theorem 2.3 an octagon, which has the same total number of vertices, always has six triangles in a triangulation, not eight.

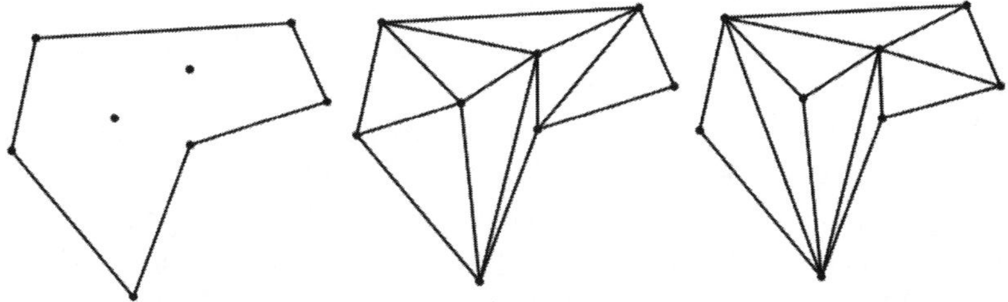

Figure 2.14. Triangulations of a hexagon with interior points.

Problem 2.4 (a). Explore whether polygons with n vertices and k interior points always have the same number of triangles in a triangulation. If so, find a formula for the number of triangles in a triangulation in terms of n and k.

With two diagonals, a square has two ways to be triangulated. A convex pentagon has five triangulations, each one using the diagonals from one of its five vertices.

Problem 2.4 (b). Find the number of ways to triangulate a convex hexagon and a convex heptagon. Describe a method (or find a formula) to find the number of ways to triangulate a convex n-gon.

A polygon has the same number of vertices as edges. Also we can use triangulations to deduce that the sum of the interior angles of a polygon depends only on its number of vertices. An n-gon can be triangulated with $n-2$ triangles, each having an *angle sum* of 180°. Thus, the angle sum of an n-gon is $180°(n-2)$. The next two problems ask you to find properties of polyhedra analogous to these properties. Figure 2.15 helps to visualize bipyramids (back-to-back pyramids) and antiprisms. (Unlike a prism, the top base of an antiprism is twisted relative to the bottom base giving triangles for sides instead of rectangles.)

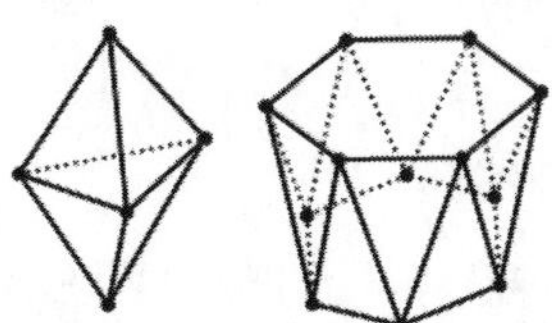

Figure 2.15. A bipyramid and an antiprism.

Remark. Actual three-dimensional models of polyhedra aid thinking much more than even the best drawings. We will consider polyhedra in a number of problems throughout this book. Consult [34] to help you build your own models.

Problem 2.4 (c). Find a relationship between the number of vertices, edges, and faces of pyramids, prisms, bipyramids, antiprisms, and general convex polyhedra.

Problem 2.4 (d). Find a relationship between the number of vertices of a convex polyhedron and the sum of all the angles of all of its faces.

The three-dimensional analog of a triangulation is to split a polyhedron into the simplest polyhedra, which are triangular pyramids or tetrahedra. ("Tetra" is Greek for four, the number of faces in a triangular pyramid.) Mathematicians coined the word *tetrahedralize* for splitting up a polyhedron into tetrahedra, with similar requirements as a triangulation splitting a polygon into triangles: the vertices of the tetrahedra must be vertices of the original polyhedron and their interiors don't overlap. Also, together the tetrahedra and their interiors cover the polyhedron and its interior, and each tetrahedron is contained in the polyhedron. Figure 2.16 illustrates one way to split a hexagonal pyramid into four tetrahedra.

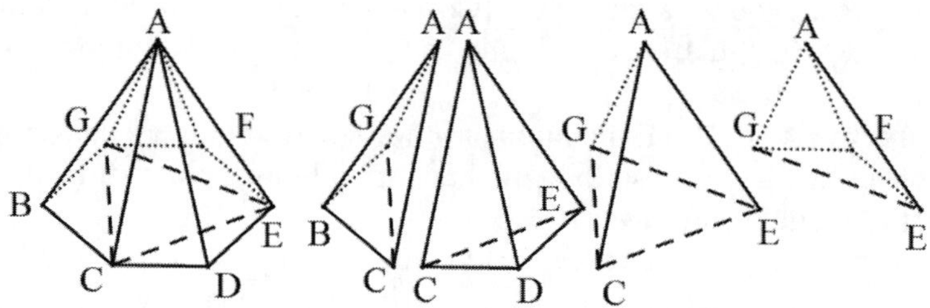

Figure 2.16. Splitting a pyramid into tetrahedra.

Problem 2.4 (e). Describe a way to tetrahedralize every pyramid. Do we always get the same number of tetrahedra?

Problem 2.4 (f). Can every convex polyhedron be tetrahedralized? If so, do we always obtain the same number of tetrahedra? If not, can we bound the minimum number of tetrahedra in terms of some aspect of the polyhedron?

2.3 Distances and Points

Problem 1.5 (a). (repeated) What is the smallest number of different (nonzero) distances n points on a line determine? What is the largest number? Can n points on a line determine any number of different distances between the smallest and largest numbers?

We can match points on a line with numbers on a number line. For ease, let the number for point i be a_i with $a_1 < a_2 < \ldots < a_n$ and write $d(x, y)$ for the distance from x to y. We have to have at least $n - 1$ different distances: $d(a_1, a_i)$ for i going from 2 to n. We can have exactly this number by choosing $a_i = i$ since the distances will be integers going from 1 to $n - 1$. (We assume from now on that the two points a_i and a_j are different since $d(a_i, a_i)$ is always 0.) Table 2.6 suggests one way to get every pair of different points to give a different nonzero distance, which will give the largest number of distances possible. We've seen the number of nonzero distances in Table 2.6 before as the triangular numbers $\binom{n}{2}$.

Table 2.6. Maximizing distances.

n	points	nonzero distances
1	1	0
2	1, 2	1
3	1, 2, 4	3
4	1, 2, 4, 8	6

The pattern in the middle column of Table 2.6 suggests one way to get all $\binom{n}{2} = \frac{n(n-1)}{2}$ nonzero distances: pick $2^{i-1} = a_i$ for the points.

There are many ways to generate all the different numbers of distances between the minimum and the maximum. For instance, start with the numbers 1 to $n - 1$ and increase the last number from n up one by one up to $2n - 2$ to get additional distances up to $2n - 1$ different distances. Next we increase the last two numbers and so on, until all but the first two numbers get spaced out to eliminate more and more duplicate distances.

Problem 1.5 (b). (repeated) Is there an arrangement of some number n of points in the plane that determine fewer distances that the smallest number of distances of n points in a line? What is the smallest number of distances 3, 4, 5, 6, or 7 points in the plane can determine?

Figure 2.17 uses regular polygons to give the minimal number of nonzero distances of various numbers of points in the plane. Different types of lines indicate different nonzero distances. Table 2.7 gives these minimal numbers of distances.

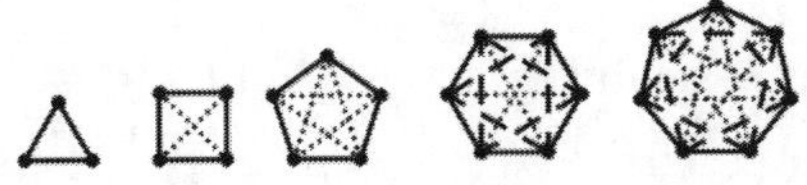

Figure 2.17. Arrangements with minimal number of distances.

Table 2.7. Minimal numbers of distances in the plane.

n	3	4	5	6	7	8
minimum number of distances	1	2	2	3	3	4

Figure 2.17 and Table 2.7 suggest that the minimal number of distances in the plane occurs with regular n-gons. Also, the minimum number of distances with n points is one half the number of vertices or a bit less when n is odd. To avoid having different cases for even and odd numbers, we can use the *greatest integer function* $\lfloor\ \rfloor$, also called the *floor function*, defined for a real number r as $\lfloor r \rfloor$ is the greatest integer less than or equal to r. For example, $\lfloor \pi \rfloor = 3 = \lfloor 3 \rfloor = \lfloor 3.97 \rfloor$.

The number of nonzero distances of a regular n-gon is then $\lfloor \frac{n}{2} \rfloor$, the greatest integer less than or equal to $\frac{n}{2}$. For $n > 2$, this value is definitely smaller than $n - 1$, the minimum number for n points on a line. But is it the minimum in the plane for bigger values of n?

Problem 2.5 (a). For some n is there an arrangement of n points in the plane with fewer distances than $\lfloor \frac{n}{2} \rfloor$, the number for a regular n-gon?

Problem 2.5 (b). Find arrangements of points in three dimensions with fewer distances than can be done in two dimensions with the same number of points.

We can find a more radical way to "question the answer." What happens if we use an alternative meaning of distance than the usual formula? For points (p, q) and (r, s) the usual (Euclidean) distance is based on the Pythagorean theorem:
$$d((p, q), (r, s)) = \sqrt{(r - p)^2 + (s - q)^2}.$$

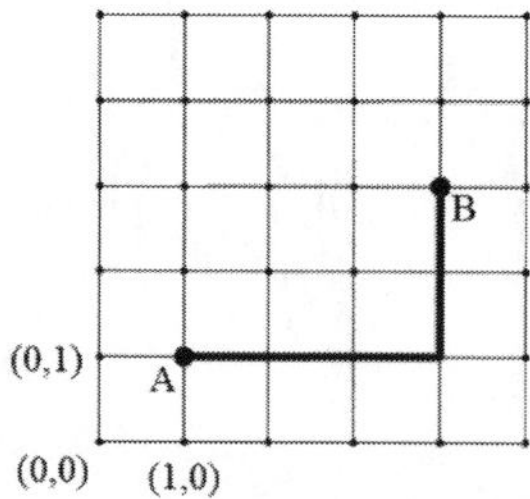

Figure 2.18. The taxicab distance from A $=$ $(1, 1)$ to B $=$ $(4, 3)$ is $d_T(A, B) = |4 - 1| + |3 - 1| = 5$.

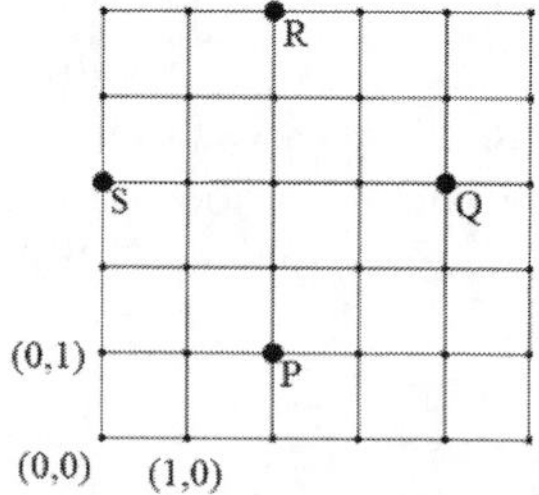

Figure 2.19. The points P, Q, R, and S are all a taxicab distance of 4 apart.

In a city with a grid of parallel and perpendicular streets, distance "as the crow flies" would not be all that helpful for those of us who can't fly. Figure 2.18 gives an example of distances using what is often called the "taxicab metric": how far is it between two places if we can only go along horizontal and vertical paths? We measure distances using the horizontal difference in the points plus the vertical difference—no complicated squaring and square rooting. Figure 2.19 gives four points in the plane having just one nonzero distance and so fewer distances than is possible with the Euclidean distance.

Definition. For points (p, q) and (r, s) the *taxicab distance* is

$$d_T((p, q), (r, s)) = |r - p| + |s - q|.$$

Problem 2.5 (c). Find the minimum number of nonzero distances of n points in the plane using taxicab distance, for $n = 5, 6, 7$, and 8.

Problem 2.5 (d). Make a conjecture about the maximum number of points in the plane that have a total of 3, 4, or 5 nonzero taxicab distances.

Problem 2.5 (e). Generalize taxicab distance to three dimensions and repeat Problem 2.5 (d) for three dimensions.

Circle packing switches the focus with regard to points (now centers of circles) and distances (radii of circles). We pack *unit circles* (of radius 1) as tightly as possible without their interiors overlapping.

Problem 2.5 (f). Find an arrangement of unit circles that you think is packed as tightly as possible. Define some way to measure how tightly packed your arrangement is. Find another packing of unit circles that has a different measure than the first one, yet doesn't have any "holes" big enough to allow any more unit circles in the region.

2.4 The Art Gallery Problem

Victor Klee (1925–2007) posed the art gallery problem in 1973 about how many guards are needed in an n-sided gallery. Two years later Václav Chvátal (1946–) found and proved the value of g_n, the maximum number of guard needed to be able to observe every point in every n-gon. Our proof of Theorem 2.5 follows the clever idea found by Steve Fisk (1946–2010) in 1978.

Problem 1.6. (repeated) Explain why $g_4 = 1$, even though there are nonconvex quadrilaterals. Find g_5. Find a polygon with the fewest number of vertices needing two guard points. Repeat for three or more needed guard points. Look for a pattern for g_n.

For a polygon to need just one guard point, that point must lie in the intersection of all the interior angles of the polygon, as pictured in Figure 2.20. Equivalently, a second guard point becomes necessary when there are two angles with no intersection within the polygon. A few sketches should convince you that $g_4 = 1 = g_5$. Figure 2.21 gives examples showing $g_6 = 2$ and $g_9 = 3$. As

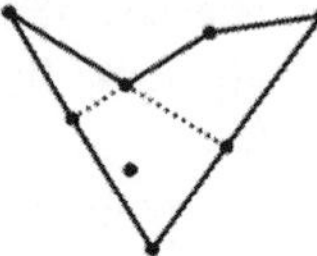

Figure 2.20. Polygon with one guard point.

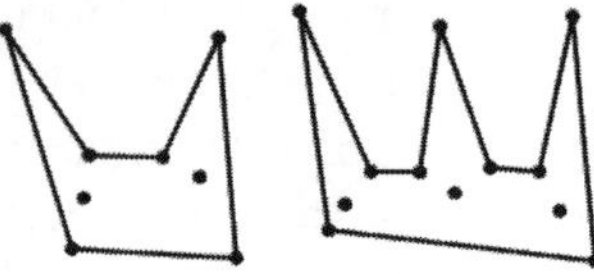

Figure 2.21. Polygons needing two and three guard points.

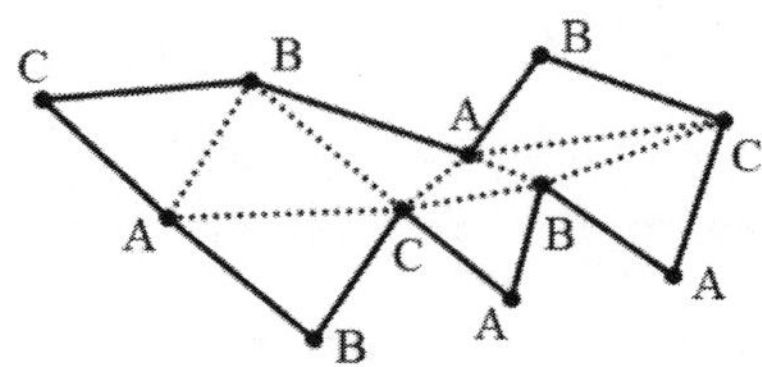

Figure 2.22. Triangulated polygon with labelled vertices. The three C's can be guard points, although only two of them are required.

Theorem 2.5 will prove, the number of guards needed is at most one third of the number of vertices. In the language of the greatest integer function, $g_n = \left\lfloor \frac{n}{3} \right\rfloor$. We can use the idea of the spikes of the polygons in Figure 2.21 to see that there are polygons with $3k$ vertices that need k guards. The proof depends on triangulating the polygon and labeling the vertices with three letters so that each triangle has all three letters. See Figure 2.22.

Theorem 2.5. (Chvátal, 1975, Fisk, 1978) The interior of any n-gon never needs more than $\left\lfloor \frac{n}{3} \right\rfloor$ guard points.

Proof by induction. Triangulate the polygon. Idea: label each vertex of the polygon with one of the letters A, B, and C so that each triangle in the triangulation has all three letters. We'll first use induction to show that such a labeling is always possible. Then we can use the vertices with the same label for our guard points.

For the base case, a triangle can have its three vertices labelled with three different letters. Any one of them can be the guard, and $\left\lfloor \frac{3}{3} \right\rfloor = 1$, showing the formula holds for the base case. For the induction step, suppose that we can appropriately label the vertices of a polygon with n vertices, where $3 \leq n$ and consider a triangulation of a polygon with $n + 1$ vertices. From Corollary 2.4, at least two of the triangles in the triangulation are ears. Let $\triangle PQR$ be an ear with $\overline{QR}$ a diagonal of the polygon. Then $\overline{QR}$ separates P from the rest of the polygon. This rest of the polygon has n vertices and so by assumption can be labelled with the three letters A, B, and C. Do so. Then Q and R have two of those three letters, leaving the third one for P. So, this labeling can be extended to all polygons with $n + 1$ vertices. By induction, we can label any polygon.

Every triangle in the triangulation has a vertex of each label and all triangles are convex. Then every point can be seen from the vertices labelled A or those labelled B or those labelled C. We pick the set with the fewest elements as the guard points. If the polygon has n vertices, then that set of guard points has at most $\left\lfloor \frac{n}{3} \right\rfloor$ vertices, completing the proof. $\qquad\square$

Instead of guarding the inside of the polygon, we could think of the polygon outlining a fortress with the guards on the boundary, and they need to see all outside points without looking through the edges (fortress walls).

Definition. A set of points $\{F_1, F_2, \ldots, F_k\}$ is a set of *fortress points* for a polygon P provided each F_i is in P and for every point A not inside P there is a fortress point F_i so that the only point of $\overline{AF_i}$ in P is F_i.

Problem 2.6 (a). Find the maximum number of points needed to form a set of fortress points for a convex polygon with n vertices. Generalize to nonconvex polygons.

Here is another twist: We can consider guards for a prison, who presumably need to be able to see every point both inside and outside the prison. (Our prison will be very simplistic: one big room surrounded by a wall with guards on the top of the wall, looking both in and out, but not through walls.)

Problem 2.6 (b). Define what a set of prison guard points is. Find the maximum number of points needed to form a set of prison guard points for a convex polygon with n vertices. Generalize to nonconvex polygons.

2.5 Geometric Patterns

We'll continue exploring the aspect of patterns introduced in Chapter 1 and venture into others as well.

Problem 1.7 (a). (repeated) Design colored patterns for regular polygons with different numbers of vertices exhibiting regularity with two or more colors. What possible numbers of colors can such a coloring have in an n-gon?

Problem 1.7 (b). (repeated) Describe what you mean by a pattern having a regular coloring.

Here is a plausible first attempt in defining color regularity: all colored regions are congruent and there is the same number of regions of one color as of any other color. We need to explore this or any proposed definition. First of all, do designs that you think look acceptable satisfy this definition? (The designs in Figure 1.6 do satisfy these conditions.) Second, can you design something that fits this proposed definition but violates your intuition of coloring regularity? Mathematicians call such an example a *counterexample*—it counters some

Figure 1.7. (repeated) Design with irregular coloring.

aspect. While the design in Figure 1.7 (repeated here) satisfies these conditions, it violates my intuition about regularity, so I want a definition to include extra conditions.

While there are four congruent regions of each color in Figure 1.7, their spacing and orientation are irregular. For instance, the four striped regions are rotations of one another, but not all the same angle each time. Three of the black regions are rotations of one another, but the other one is flipped. Also, the spacing for the black regions is irregular. The same thing holds for the white regions as the black ones.

Let's use rotations and reflections (flips over lines) to describe the regularity of the designs of Figure 1.6 (reproduced here). In the first and third designs, rotations of 120°, 240°, and (of course) 0° = 360° take regions of one of the two colors to regions of the same color. We say these rotations *preserve colors*. Rotations of 60°, 180°, and 300° *switch colors*. For the second design, rotations of multiples of 60° preserve colors, while any of six reflections between adjacent regions switch colors. For the first and third designs, three reflections preserve colors and three switch colors, although which reflections do the preserving or switching differs.

The three-colored design at the right of Figure 1.6 has more complicated interactions of the colors. Rotations of 180° and 0° preserve colors, rotations of 60° and 240° take black to striped to white back to black, and rotations of 120° and 300° reverse that shift of colors. Reflections over a vertical or a horizontal line preserve the black color and switch the other two. Similarly, there are two diagonal reflections preserving the white color and switching the others and two other reflections preserving the striped color and switching the others. The rotations

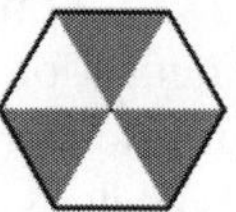 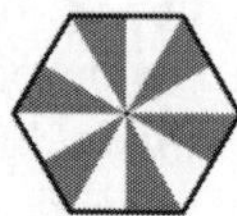 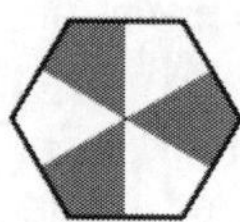 

Figure 1.6. (repeated) Regular colorings of a hexagon.

and reflections are the *symmetries* of the design; that is, transformations that take the design to itself and keep the organization of the design.

For Figure 1.7, reflections and nonzero rotations mess with the organization of the colors. For instance, a rotation of 180° takes two of the black triangles to black triangles, one to a white triangle, and one to a striped triangle. Only the rotation of 0° preserves colors. In my understanding of regularity for a given symmetry, all regions of color x are either preserved or all go to regions of some other color y. Also, if A and B are regions of the same color, then there is a symmetry taking A to B. Actually, even if the regions are different colors, I want some symmetry to map one to the other, while shifting all the colors compatibly. The definitions of symmetry and regular coloring capture all of these properties.

Definition. A *symmetry* of a design with colored regions is a function that maps the design onto itself so that each region is mapped to another region of the design, and if two regions have the same color before the mapping, they are mapped to regions with the same color.

Definition. A design with colored regions has a *regular coloring* provided for any two regions A and B there is a symmetry taking A to B.

To answer Problem 1.7 (a) using these definitions, we need to know the symmetries that designs based on regular polygons can have. We have already discussed the possibilities, as Leonardo da Vinci (1452–1519) recognized over 500 years ago. The mathematics to prove the statement in Lemma 2.5 developed over several hundred years after da Vinci's death. We will follow da Vinci and take Lemma 2.5 as given, rather than develop the needed geometry to prove it. (Mathematicians call a theorem a lemma if it is seen as only a stepping stone toward proving a theorem.)

Lemma 2.5. The possible symmetries of a regular n-gon are rotations of the n multiples of $\frac{360}{n}^\circ$ and n reflections over lines through the center.

Proof. See Figure 2.23 and [29, 268]. $\square$

The color-preserving and color-switching symmetries of the designs of Figure 1.6 all come from the twelve possibilities in Figure 2.23, where R^i is a rotation of $\frac{360}{6}i^\circ$. We can compose two of these symmetries together to get another symmetry. We use $\circ$ for the operation of composing symmetries or functions. For example,

$$R^2 \circ R^3 = R^5 \quad \text{and} \quad M_3 \circ M_1 = R^2.$$

The composition in the last equation may seem backward because it follows the function notation: $f \circ g(x) = f(g(x))$. For $M_3 \circ M_1 = R^2$, we first apply reflection M_1 to points and then apply M_3 to where those points end up. So the midpoint of the edge on the top of the hexagon stays where it is when we apply M_1. Then M_3 takes it to the midpoint of the lower left edge of the hexagon. This

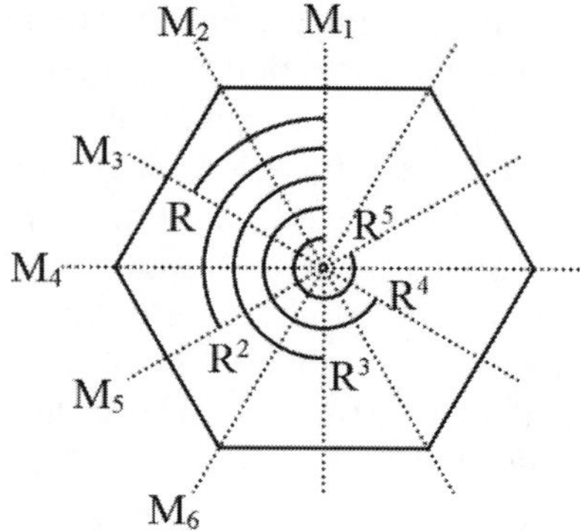

Figure 2.23. Rotations and (mirror) reflections of a hexagon.

is equivalent to rotating by 120°, what R^2 does. The composition $M_3 \circ M_1$ does this rotation for every other point as well.

Theorem 2.6 will give a strong relationship between the symmetries, regions, and colors of a design in a regular polygon. A given design based on an n-gon need not have all $2n$ symmetries, as Figure 2.24 illustrates. Its only color-preserving symmetry is the identity, the rotation of 0°. Rotations of 120° and 240° shift the three colors around. The three reflections over lines separating two regions switch the colors of those two regions and leave the other region its original color.

The set of color-preserving symmetries and the entire set of symmetries form abstract algebraic systems called *groups*. We'll call them the *color-preserving group* and the *color group*. Over the last two centuries group theory has become a central idea in mathematics. We don't need the formalism of group theory, but the concept of a group and Theorem 2.6 will prove useful.

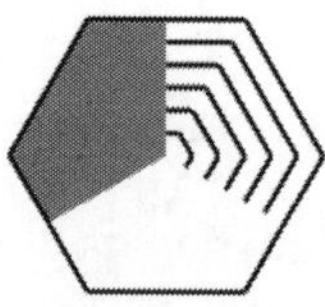

Figure 2.24. A regular coloring of the hexagon with one color-preserving symmetry and six symmetries total.

Definition. A set G with an operation $\circ$ is a *group* provided

(a) for all g and h in G, $g \circ h$ is in G;

(b) there is an *identity* element e of G so that for all g in G, $g \circ e = e \circ g = g$;

(c) for each g in G there is an *inverse* g^{-1} in G so that $g \circ g^{-1} = g^{-1} \circ g = e$; and

(d) for all g, h, and j in G, $g \circ (h \circ j) = (g \circ h) \circ j$.

Our familiar set of numbers forms a group with addition as the operation. The identity is 0 and the additive inverse of x is $-x$. The positive numbers form a group using multiplication with identity 1 and $\frac{1}{x}$ as the multiplicative inverse of x. (We could include negative numbers for this second group, but not 0 since 0 has no multiplicative inverse.)

More importantly for our purpose, the symmetries of a design form a group using composition of functions. The identity is the transformation that doesn't move anything—a rotation of $0°$. The inverse of a rotation of $x°$ is a rotation of $(360-x)°$, and the inverse of a reflection is itself. Since the color-preserving group of a design is a subset of the color group of the design and a group on its own, we say it is a *subgroup* of the color group. Theorem 2.6 relates the number of colors, regions, symmetries, and color-preserving symmetries for designs with a finite number of symmetries. It is a special case of Lagrange's theorem, a key insight in group theory. (See [30, 87] for Lagrange's theorem.)

Theorem 2.6. Suppose that a design with regular coloring has a finite number of regions and so finitely many symmetries. If the design has k colors, and the subgroup preserving one color has n symmetries, then the color group has nk elements. Also, the number of colors divides the number of regions, which divides the number of symmetries.

Before we prove this theorem, let's see it in action with the designs in Figure 1.6. All the designs have twelve symmetries in their color group. The first, third, and fourth designs have six regions, while the second one has twelve regions. The number of colors, two or three, divides the number of regions. For the first, second, and third designs, there are two colors, six symmetries preserving each color, and $2 \times 6 = 12$ symmetries altogether. For the fourth design, there are three colors and just one symmetry preserving all colors, but there are four symmetries preserving any one color. And 12, the size of the color group, equals 3×4. (The four symmetries preserving the color white differ from the four preserving black and both of these differ from the four preserving stripes. When there are just two colors, a symmetry preserving one of the colors necessarily preserves the other color.)

Proof. Suppose the symmetries preserving color c_1 are $s_1, s_2, \ldots, s_n$. Also, since the coloring is regular, we have particular symmetries $g_2, g_3, \ldots, g_k$, where

g_i takes regions of color c_1 to color c_i. For j with $1 \leq j \leq n$, the composition of symmetries $g_i \circ s_j$ takes any region of color c_1 to a region with color c_i, giving n symmetries taking color c_1 to color c_i. With k colors we have the desired number of nk candidates for symmetries. Exercises 2.1 and 2.2 will show that all nk candidates differ from one another and that there are no others. (Answers to these and other exercises appear after Chapter 4.)

Similarly, suppose that there are r regions, $R_1, R_2, \ldots, R_r$. Since the coloring is regular, for each i, there is a symmetry h_i taking region R_1 to R_i. If there are z regions the same color as R_1, there are z regions the color of any R_i. With k colors, there must be kr regions, showing that the number k of colors divides the number of regions.

Finally, we show that the number of regions r divides the number of symmetries. Suppose j symmetries take the region R_1 to itself. We use an argument similar to the first five sentences of the proof to show that there are j of these symmetries taking R_1 to each of the other regions. This means that there are jr symmetries altogether, and so r divides the number of symmetries. $\qquad\square$

Exercise 2.1. First we show that there are at least nk symmetries. Let $g_i \circ s_j$ and $g_t \circ s_u$ be two candidates, with $1 \leq i \leq k$, $1 \leq t \leq k$, $1 \leq j \leq n$, and $1 \leq u \leq n$. We assume that $g_i \circ s_j = g_t \circ s_u$, and we need to prove that $i = t$ and $j = u$ to show that all the nk candidates are different symmetries. Use colors to explain why we can't have $i \neq t$ when $g_i \circ s_j = g_t \circ s_s$. Now assume that $i = t$ and so $g_i \circ s_j = g_i \circ s_u$. Use the inverse g_i^{-1} to deduce $s_j = s_u$. Thus we have at least nk symmetries.

Exercise 2.2. To show that there are no other symmetries, let h be any symmetry. It must take color c_1 to some color, say c_i. If h takes color c_1 to itself, it is one of the s_j. For $i > 1$, the symmetry g_i also takes color c_1 to c_i. Show that $g_i^{-1} \circ h$ preserves color c_1 and so is one of the symmetries $s_1, s_2, \ldots, s_n$, say s_w. Show how to write h as a composition of g_i and that s_w.

With Theorem 2.6 we can answer Problem 1.7 (a): the number of colors in a design must divide the number of symmetries in the design.

It's time to question the answer. While Figure 1.7 wasn't a regular coloring of the hexagon, at least adjacent regions were different colors. That way, we can tell when two regions are different. Since 1852, mathematicians have investigated coloring maps using the fewest number of colors—where regions of a map are required to have different colors whenever they share a boundary. (But regions sharing only a vertex, like diagonal squares on a checkerboard, can have the same color.) We won't try to define maps or colorings rigorously. For ease we will only consider regions that are polygons with adjacent regions sharing an edge. Figure 2.25 shows several maps requiring different numbers of colors. In 1852, Francis Guthrie (1831–1899) noticed that no map on a plane ever seemed to require more than four colors.

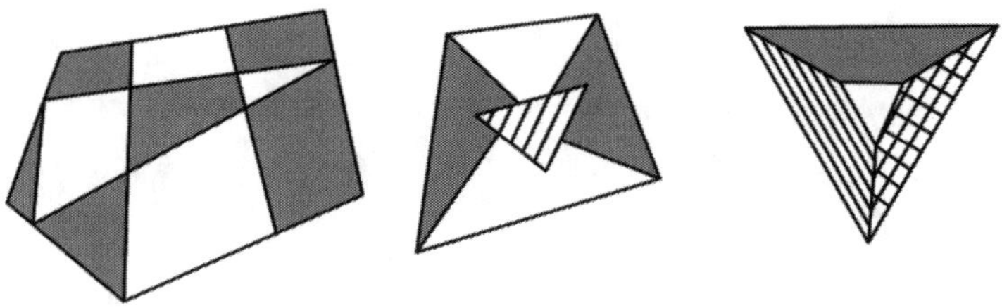

Figure 2.25. Maps requiring two, three, and four colors.

The four-color conjecture. Every map on a plane can have its regions colored using at most four colors so that no adjacent regions are the same color.

This conjecture engaged lots of mathematicians and generated many false proofs over the following 120 years. Percy Heawood (1861–1955) did prove in 1890 that five colors were sufficient to color any map on a plane (or on a sphere). It also engendered the development of much sophisticated mathematics and computer programming and even some applications. (See [5] and [36].)

Problem 2.7 (a). Draw a number of maps and color them with the minimum number of colors so that no adjacent regions are the same color. Find conditions when the map requires at least four colors and when it can be colored with just two colors.

Another way to question the answer of Problem 1.7 is by investigating infinite regular patterns instead of finite ones.

Artists have made friezes (decorations along a horizontal band, often just below the ceiling of a room) for thousands of years. Mathematicians require a *frieze pattern* to be a design repeated infinitely many times along a line in set intervals so that translations are symmetries for the entire pattern. For ease, we will assume the translations are always horizontal. Figures 2.26 and 2.27 both illustrate (parts of) two frieze patterns, the left one of each having only translations for symmetries and the right one having 180° rotations as well as translations. I trust you find the Mexican friezes more interesting than the ones made with letters.

$$\ldots\, \text{F F F F F} \,\ldots \qquad\qquad \ldots\, \text{Z Z Z Z Z} \,\ldots$$

Figure 2.26. Two frieze patterns made with letters.

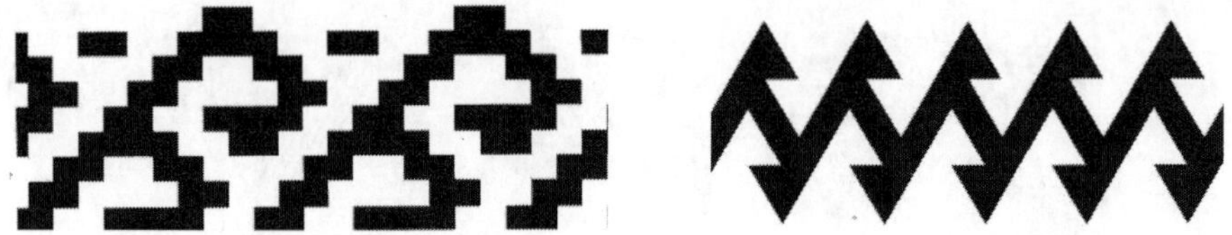

Figure 2.27. Two Mexican frieze patterns of the same symmetry types as Figure 2.26.

Problem 2.7 (b). Design as many different types of frieze patterns as you can, where two types are different when one has a symmetry the other doesn't have.

It will help you do Problems 2.7 (b) and later (c) if you know what the possible symmetries are. By definition, each design must have horizontal translations (abbreviated T). From Figures 2.26 and 2.27, rotations of 180° (R) are also possible. The other options are horizontal mirrors (H), vertical mirrors (V), and horizontal glide reflections (G). The transformation that takes triangle A in Figure 2.28 to triangle B is a glide reflection. It can be built from a translation (glide) taking triangle A to the lightly shaded triangle followed by the horizontal mirror flipping that triangle over to triangle B. (See [29, 273–274] for a proof that these are the only symmetries of a frieze pattern.)

We can denote the two types of frieze patterns in Figures 2.26 (and 2.27) by T and TR, respectively. (People who study patterns more formally have other names. See [29, 275].) Problem 2.7 (b) thus asks you to find other frieze types that include T and some or all of the other options V, H, R, and G.

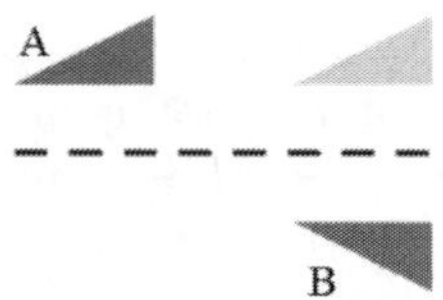

Figure 2.28. A glide reflection taking triangle A to triangle B.

We can make much more interesting frieze patterns if we introduce regular colorings. As with other two-color patterns, there are two groups: the color group of all symmetries and the subgroup of color-preserving symmetries. The color-preserving group for the two-color frieze pattern in Figure 2.29 has only translations. The color group has in addition 180° rotations. We can denote this pattern as TR/T, where we list the color group first and the color-preserving group second. For the two-color frieze pattern in Figure 2.30, both the color-preserving

Figure 2.29. A two-color frieze pattern of type TR/T.

Figure 2.30. A two-color frieze pattern of type TV/TV.

subgroup and the entire color group have translations and vertical reflections. We denote this pattern as TV/TV. (There are shorter translations and more frequent reflections in the entire color group than in the color-preserving subgroup.)

Problem 2.7 (c). Investigate two-color frieze patterns. Look for as many different types as you can, listing the types of all symmetries and the types of color-preserving symmetries. Hint: There are seventeen possible types of two-color frieze patterns.

Tessellations (filling the plane with one or a few repeated shapes) and their colorings provide another way to investigate patterns. Figure 2.31 suggests that copies of any triangle can fit together to fill the plane with no gaps or overlaps. The figure also provides a regular coloring of the pattern with two colors.

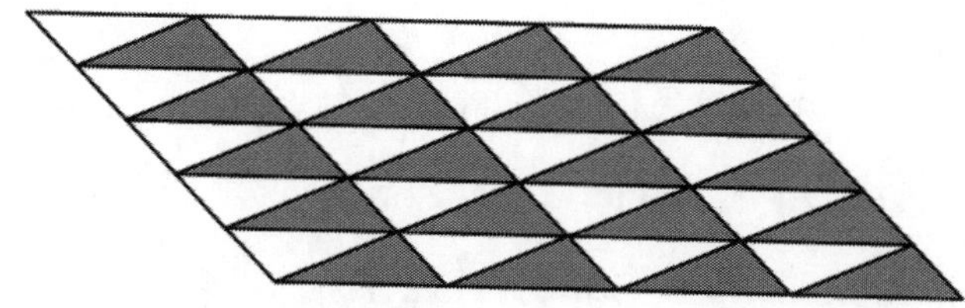

Figure 2.31. A regular two-coloring of a tessellation with copies of a triangle.

Problem 2.7 (d). Is there a regular three-coloring and a regular four-coloring of the tessellation in Figure 2.31?

Problem 2.7 (e). Make tessellations and regular colorings using each of the quadrilaterals in Figure 2.32.

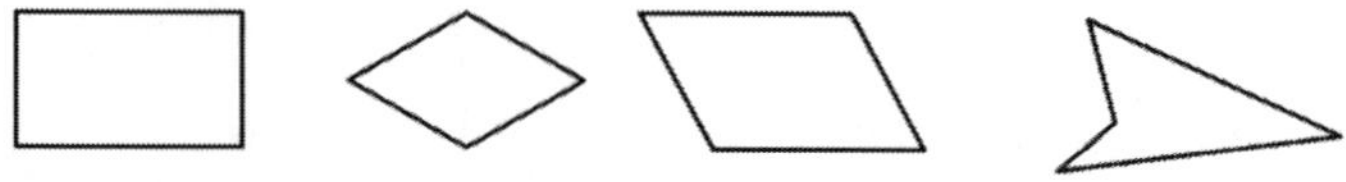

Figure 2.32. Templates for tessellations: a rectangle, a rhombus, a parallelogram, and a general quadrilateral.

2.6 Voronoi Diagrams

Problem 1.8. (repeated) Find Voronoi diagrams when there are two, three, and four sites. What property or properties in terms of the given sites characterize the line segments and rays dividing the regions of a Voronoi diagram?

The points equidistant from two given sites lie on their perpendicular bisector. In Figure 2.33, M is the midpoint of A and B and k is the perpendicular bisector of $\overline{AB}$. Points on the same side of k as A are closer to A than to B. With just two sites, a Voronoi diagram looks like Figure 2.33. In general, a Voronoi diagram has regions whose boundaries are built from parts of some or all of the perpendicular bisectors of each pair of sites. With three or more sites, such as C, D, and E in Figure 2.34, the rays (or segments) can have a common point equidistant from several sites, which is point F in Figure 2.34. Unless the original three

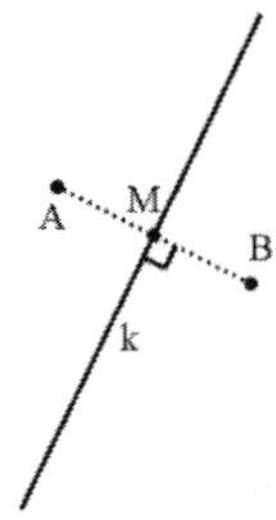

Figure 2.33. M is midpoint, k is the perpendicular bisector of $\overline{AB}$.

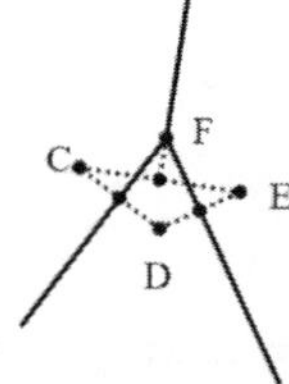

Figure 2.34. F is equidistant from C, D, and E.

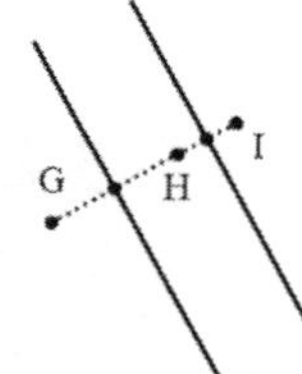

Figure 2.35. G, H, and I are collinear.

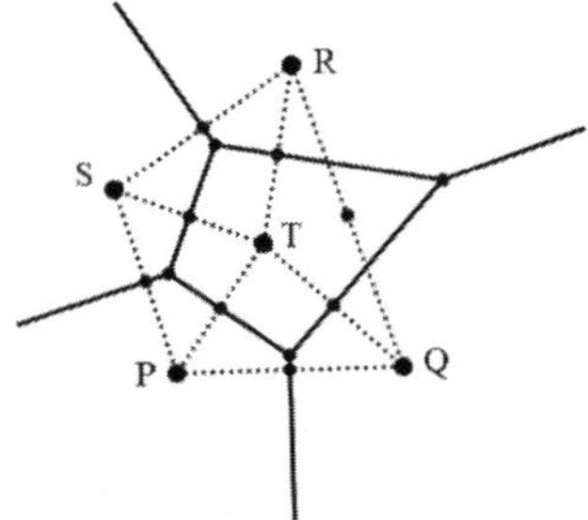

Figure 2.36. The Voronoi diagram for P, Q, R, S, and T.

sites are on a line, as in Figure 2.35, Voronoi diagrams for three sites look much like Figure 2.34. With four or more sites it is possible to have finite regions, as in Figure 2.36. Note that in Figure 2.36 no part of the perpendicular bisectors of $\overline{PR}$ and $\overline{QS}$ appears in the Voronoi diagram since the other sites come between them. The shapes of the Voronoi regions in Figures 2.33–2.36 are all convex, which suggests our next problem.

Problem 2.8 (a). Is every region of a Voronoi diagram convex?

Problem 2.8 (b). What is the shape of the Voronoi regions if the sites form a repeating lattice with translations in different directions, as in Figure 2.37?

Figure 2.37. A repeating lattice of sites for a Voronoi diagram.

Problem 2.8 (c). Devise a method to find the Voronoi diagram for a set of n sites in the Euclidean plane.

3

Further Variations

3.1 Lines and Regions

Problem 2.1 (a). (repeated) By using some placement besides general position or all parallel lines, can we have any number of regions between the minimum and maximum number of regions in the plane determined by n lines?

Some drawings, like Figure 3.1 for three lines, may convince you that we can't have every number of regions between $n + 1 = 4$ and $\frac{n^2+n+2}{2} = 7$. But we need an argument, not just a picture. Focus on the intersections of the lines in Figure 3.1, rather than the regions. The number of intersections ranges from zero to three. Zero intersections corresponds to parallel lines and the minimum number of regions, four. For three lines to have only one intersection, they all go through the same point, making six regions. To have exactly two intersections, two of the lines must be parallel and the third line intersects both. In this case, we also get six regions. With the maximum number of intersections $(3 = \binom{3}{2})$ we get the maximum number of regions (seven).

For any number n of lines, with zero intersections the n lines are parallel, and we get the minimum number of regions, $n + 1$. In all other cases, there are at

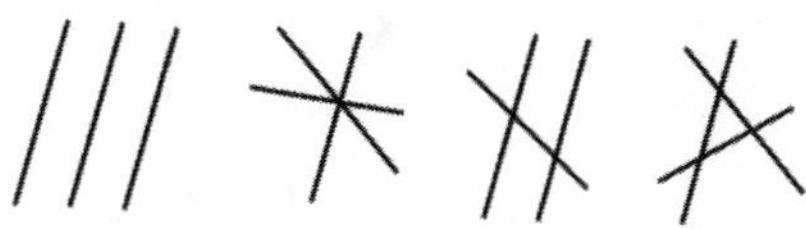

Figure 3.1. Three lines can determine 4, 6, or 7 regions.

least two lines with one intersection which already create four regions. Each line we draw cuts at least one of these two lines, creating at least two more regions. So with at least one intersection and n lines, there are at least $2n$ regions.

Problem 3.1 (a). Try to find arrangements of 4, 5, and other values of n lines to get all the values for the number of regions they determine between $2n$ and $\frac{n^2+n+2}{2}$.

Problem 2.1 (b) introduced the challenge of three-dimensional visualization with its difficulty of drawing helpful figures.

Problem 2.1 (b). (repeated) What are the maximum and minimum number of regions in three-dimensional space determined by n planes?

As with the two-dimensional version, parallel planes give the minimum number of regions, $n+1$ regions for n planes. To determine the maximum we need to understand what we should mean by general position in three dimensions. Two nonparallel planes intersect in a line. A third plane not parallel to either of the others has at least one point of common intersection with both of them. It is possible, but not in general position, for the three planes to have a common line of intersection. Figure 3.2 illustrates three planes in general position and the eight regions they determine, four above the horizontal plane and four below it.

It is hard to visualize or draw a fourth plane in general position, but we can say it will intersect each pair of the planes in a single point, illustrated in Figure 3.2 by the large dots. In Figure 3.3, we see the intersections of these three planes on the fourth plane. Each of the first three planes intersects the fourth plane in a line. These three lines are in general position, so they split the fourth plane into seven regions. Each of these seven flat regions splits one of the three-dimensional regions. That is, four planes in general position give us $8 + 7 = 15$ regions. This reasoning suggests how to continue adding more planes. We look at the regions formed in the new plane as it intersects all the previous ones. Our

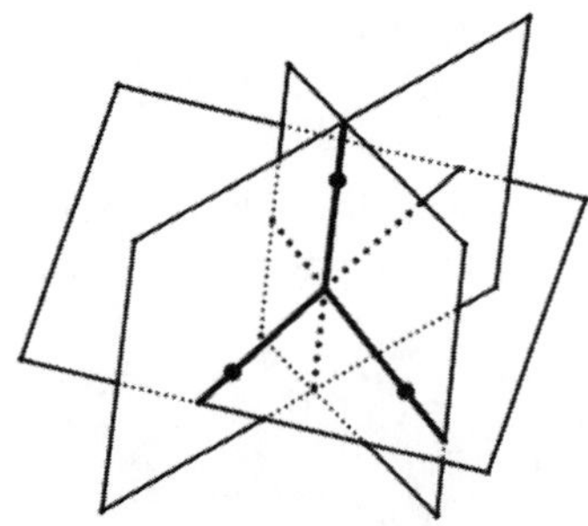

Figure 3.2. Three planes and their eight regions.

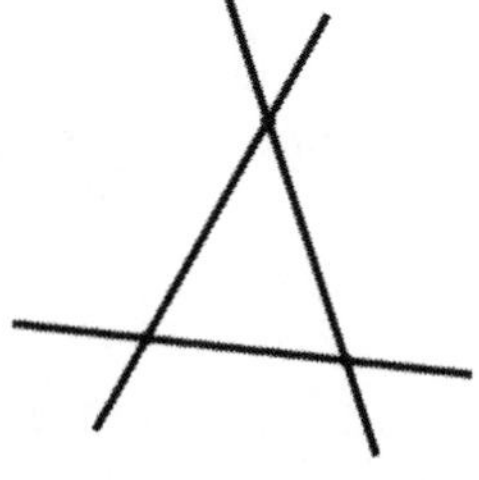

Figure 3.3. A fourth plane's intersection with three planes.

solution to Problem 1.1 tells us the number of regions the new plane is divided into. These flat regions split the same number of three-dimensional regions. The total number of three-dimensional regions is the sum of the new and old regions.

Table 3.1 gives numerical data for this reasoning. The top line is the number n of planes. The middle line gives the number of two-dimensional regions the last plane is divided into by the previous $n - 1$ planes, all in general position. This is the maximum number of plane regions, $M(n - 1)$ determined by $n - 1$ lines. The bottom line gives the maximum number of three-dimensional regions determined by n planes. Each bottom number is the sum of the number above it and the number to its left.

Table 3.1. The maximum number of regions with n planes.

n planes	1	2	3	4	5	6
2-d regions		2	4	7	11	16
3-d regions	2	4	8	15	26	42

The numbers in the top line follow the easiest first-degree expression in n, just n. We know the numbers in the second line satisfy a second-degree expression, namely $\frac{1}{2}n^2 - \frac{1}{2}n + 1$. You might guess that the numbers in the third line satisfy a third degree expression. Let's call this value $M_3(n)$. Some algebra will allow us to find the values of a, b, c, and d in $M_3(n) = an^3 + bn^2 + cn + d$. We put in four values of n, say 1, 2, 3, and 4 to get the four equations below in the four unknowns a, b, c, and d.

$$\begin{cases} 1a + 1b + 1c + d = 2 \\ 8a + 4b + 2c + d = 4 \\ 27a + 9b + 3c + d = 8 \\ 64a + 16b + 4c + d = 15 \end{cases}$$

We can solve these equations algebraically to find the only solution of $a = \frac{1}{6}$, $b = 0$, $c = \frac{5}{6}$, and $d = 1$. You can check that $M_3(n) = \frac{1}{6}n^3 + \frac{5}{6}n + 1$ gives the correct values not only for $n = 1$ to $n = 4$, but also for $n = 5$ and $n = 6$ from Table 3.1. Exercise 3.2 after the answer to Problem 2.1 (c) uses difference equations to show that this formula holds for all n.

One of the wonderful parts of mathematics is the freedom to go beyond what we can concretely visualize. In the process, our geometric intuition expands. Mathematicians have productively investigated geometry in more than three dimensions for over 180 years. We will work here by analogy, although to prove results we would need careful definitions. We consider a point to be a zero-dimensional object. More familiarly, a line is one-dimensional and a plane is

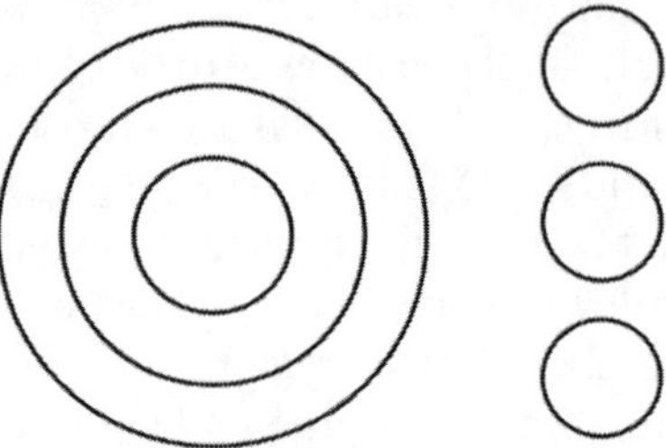

Figure 3.4. Minimum number of regions with three circles.

two-dimensional. By analogy there are three-dimensional and higher dimensional objects mathematicians call *hyperplanes* when they think of them as embedded as a part of a space of more dimensions. In this section we are dividing a space into regions. A point (zero dimensions) divides a line (one dimension) into two regions, a line (one dimension) divides a plane (two dimensions) into two regions, and a plane (two dimensions) divides three-dimensional space into two regions. By analogy, an n-dimensional hyperplane divides an $(n+1)$-dimensional space into two regions. You will need to develop the idea of general position in higher dimensions to work on the next problem.

Problem 3.1 (b). What are the maximum and minimum number of regions in four-dimensional space determined by n three-dimensional hyperplanes? Generalize for five-dimensional space and higher dimensions.

Problem 2.1 (c). (repeated) What are the minimum and maximum number of regions in the plane determined by n circles?

We obtain the minimum number of regions, $n+1$, when, as with lines, the n circles have no points of intersection, as seen, for example, in either arrangement of Figure 3.4.

We illustrate the four regions determined by two circles on the left of Figure 3.5 as a Venn diagram. Set A is inside the left circle and set B is inside the right circle. The region labeled A∩B represents those things in both sets, where the symbol ∩ represents the intersection of two sets. The notation A^c indicates things in the complement of A, things not in A. So, $A \cap B^c$ indicates things in A but not in B, and $A^c \cap B^c$ describes things in neither A nor B. On the right of Figure 3.5, we labeled the three circles A, B, and C but leave to you the designation of the $8 = 2^3$ regions the circles determine. There can't be more than eight regions for subsets of three sets: either A is or isn't in a region and the same for B and C. With two choices for each set, there are $2 \times 2 \times 2 = 2^3 = 8$ possible options. In your exploration, you may have noticed that we can't get $16 = 2^4$ regions using four circles. A Venn diagram with four sets would need sixteen regions to represent all the logical possibilities to be or not be in the different sets.

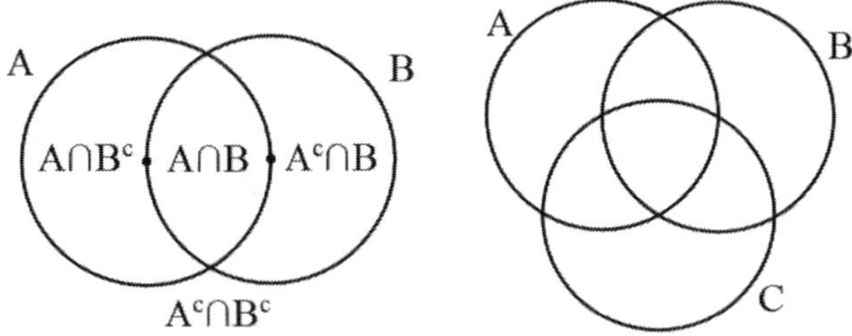

Figure 3.5. Regions determined by two and three circles.

Problem 3.1 (c). Design a Venn diagram for four convex sets that has all sixteen possible regions.

The most regions we can determine with four circles is fourteen regions, accomplished in Figure 3.6. Why no more than fourteen? Let's modify the approach we used for Problem 1.1 for regions made with lines. Two different circles can intersect in at most two points. So, in general position the fourth circle D intersects each other circle in two points, giving six points of intersection. (Figure 3.6 places dots at these intersections.) These six points cut the fourth circle into six arcs, which in turn split six of the regions formed by the circles A, B, and C. The other two regions aren't split.

The reasoning in the previous paragraph generalizes recursively to n circles. Let $C(n)$ be the maximum number of regions determined by n circles. The nth circle can intersect each of the previous $n-1$ circles in at most two points, cutting this last circle into $2n - 2$ arcs. These can split $2n - 2$ of the previous regions in two, giving a total of $C(n) = C(n - 1) + 2n - 2$ regions. We have seen recursive formulas like this several times now. We managed to turn these previous

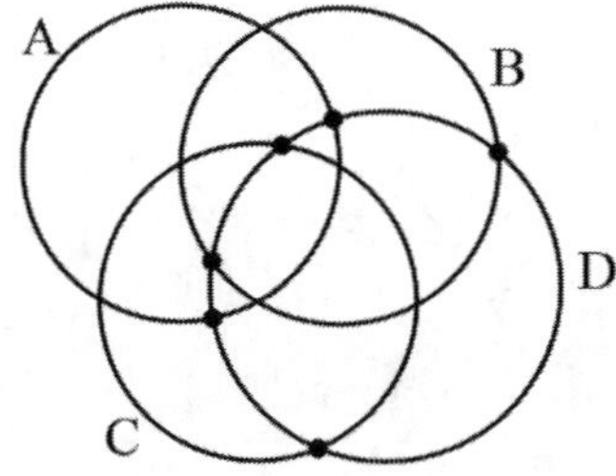

Figure 3.6. Four intersecting circles can determine fourteen regions.

recursive formulas into closed formulas by recognizing special patterns. Difference equations often provide a systematic approach to finding closed formulas without seeing special patterns.

Difference equations. We rewrite $C(n) = C(n-1)+2n-2$ as the difference of two successive terms, $C(n) - C(n - 1) = 2n - 2$, an example of a difference equation. If $f(n) - f(n - 1)$ is a formula in terms of n, we may be able to find $f(n)$ in closed form. We'll consider some of the types of differences we can get and use one of them to find the closed form for $C(n)$.

If f is a first-degree function, $f(n) = an + b$, then $f(n) - f(n-1) = an + b - (a(n-1)+b) = a$, a constant difference. So, if the function goes up by a constant difference, we know a, the coefficient of n, right away, and can determine the other coefficient from one particular value of $f(n)$. For instance, suppose that $f(1) = 3$ and $f(2) = 7$, giving a difference of $a = 4$. Then $f(n) = 4n + b$. When $n = 1$, we have $3 = f(1) = 4 \cdot 1 + b$. Then $b = -1$ and $f(n) = 4n - 1$.

If f is a second-degree function, $f(n) = an^2 + bn + c$, then $f(n) - f(n-1) = an^2 + bn + c - (a(n - 1)^2 + b(n - 1) + c) = 2an + b - a$. For Problem 2.1 (c), $C(n) - C(n - 1) = 2n - 2 = 2an + b - a$. Thus, $a = 1$ and $b = -1$. The value $C(2) = 4$ tells us $c = 2$ since $4 = 1(2^2) - 1(2) + c$. That is, $C(n) = n^2 - n + 2$.

Exercise 3.1. For a third-degree function, suppose $f(n) = an^3 + bn^2 + cn + d$. Simplify the difference $f(n) - f(n - 1)$.

Exercise 3.2. Use Exercise 3.1 to resolve Problem 2.1 (b) using a difference equation. That is, let $M_3(n)$ be the maximum number of three-dimensional regions determined by n planes. We saw that $M_3(n) = M_3(n - 1) + M(n - 1) = M_3(n - 1) + (\frac{1}{2}(n - 1)^2 + \frac{1}{2}(n - 1) + 1)$. Deduce the formula we found for $M_3(n)$.

Problem 3.1 (d). What is the maximum number of regions in space determined by n spheres?

Problem 2.1 (d). (repeated) What is the maximum number of regions in the plane determined by two convex polygons with j and k sides?

Figure 3.7 suggests that the smaller of j and k determines the maximum number of regions. The key is that a convex polygon can only cross an edge twice. For $j \le k$, the k-gon can only intersect the j-gon in $2j$ points, making a sort of

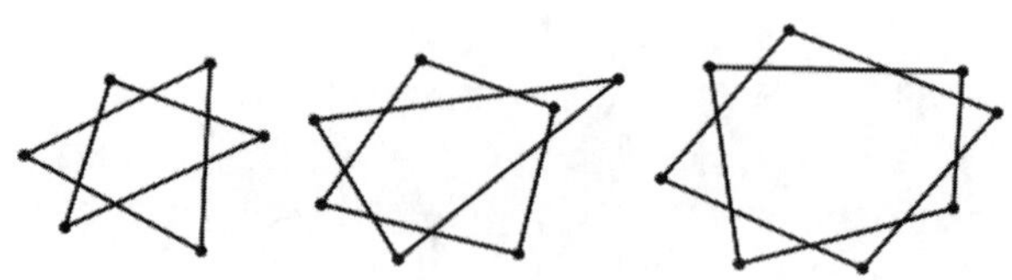

Figure 3.7. Regions determined by two convex polygons.

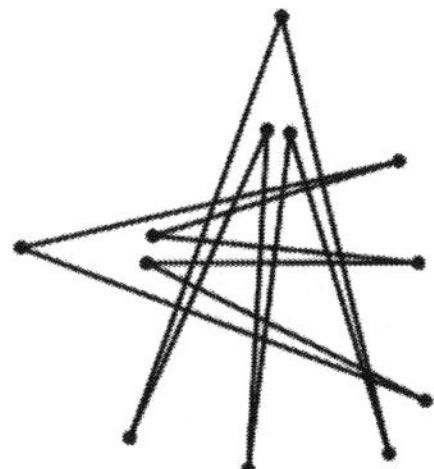

Figure 3.8. Two intersecting nonconvex polygons.

$2j$-pointed star shape. That is, there can be at most $2j$ jutting regions plus the inside and outside regions, giving at most $2j + 2$ regions.

Problem 3.1 (e). What is the maximum number of regions in the plane determined by two nonconvex polygons each with the same even numbers of sides? (See Figure 3.8.)

3.2 Diagonals and Triangulations

Problem 2.2. (repeated) Does every convex polyhedron that isn't a pyramid have an interior diagonal?

In mathematics one counterexample suffices to disprove a general statement. This differs from everyday life, where we have all heard the phrase "the exception that proves the rule." A triangular prism has no interior diagonals. As Figure 3.9 illustrates, there are nine edges, and the three rectangular faces each have two surface diagonals. That accounts for all $\binom{6}{2} = 15$ segments between the six vertices of the prism, leaving no interior diagonals.

Are there other examples? We can certainly modify the prism by squashing a bit to have parallelograms instead of rectangles or by shortening one of the edges between the triangular faces so they aren't parallel and two of the four-sided faces become trapezoids or other modifications. But these all have the basic same format: five faces of which two are triangles and three are convex quadrilaterals, illustrated in Figure 3.10. Those quadrilaterals each have two surface diagonals, for a total of six surface diagonals. These together with the nine edges account for all $\binom{6}{2} = 15$ possible segments, and so no interior diagonals. The definition below specifies what qualifies as what we will call a modified triangular prism. Theorem 3.1 shows that convex polyhedra without interior diagonals are either pyramids or modified triangular prisms. The key to eliminating other polyhedra is showing certain sets of four vertices have to be *coplanar*; that is, they lie in the same plane, and so form a quadrilateral.

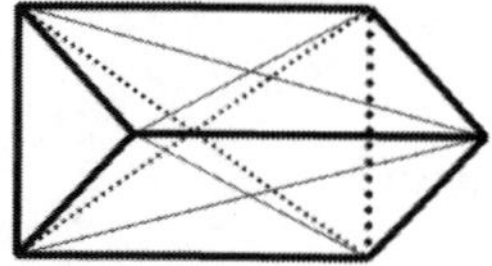

Figure 3.9. A triangular prism. Edges are thick, diagonals thin.

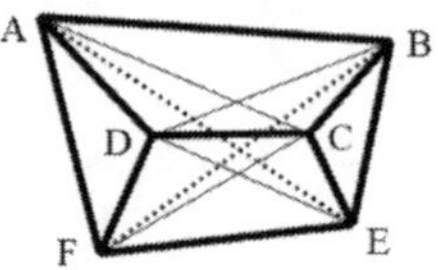

Figure 3.10. Modified triangular prism. Edges are thick, diagonals thin.

Definition. A *modified triangular prism* is a polyhedron with six vertices A, B, C, D, E, and F so that ABCD, ABEF, and CDFE are convex quadrilaterals and ADF and BCE are triangles.

Theorem 3.1. A convex polygon with no interior diagonals is either a pyramid or a modified triangular prism.

Proof. We consider four cases, starting with the case where all faces are triangles.

Case 1. In this case, where all faces are triangles, there are no surface diagonals, and so with n vertices there need to be $E = \binom{n}{2} = \frac{n(n-1)}{2}$ edges. Further, each vertex has $n - 1$ edges linking it to each of the $n - 1$ other vertices and $n - 1$ triangles sharing that vertex. The smallest value of n to give a three-dimensional shape is $n = 4$. In this case, we have a triangular pyramid with six edges and four faces. I claim that this is the only possibility. Suppose now that there are at least five vertices in a convex polyhedron, and all of its faces are triangles. We'll look for an interior diagonal. Consider a vertex A with adjacent vertices B, C, D, E, …. Then we have triangles $\triangle ABC$, $\triangle ACD$, $\triangle ADE$, …. Consider the segments $\overline{BD}$ and $\overline{CE}$. (See Figure 3.11.) They have to either both make triangular faces or at least one is an interior diagonal. The triangles would be $\triangle BCD$ and $\triangle CDE$. They can't both be triangles because they would intersect. Thus only a triangular pyramid satisfies this first case.

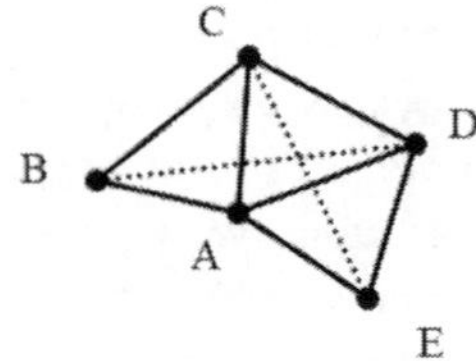

Figure 3.11. An attempted polyhedron for Case 1.

Case 2. There is an *n*-gon with $n > 3$ and exactly one vertex not in the plane of the *n*-gon. This is a pyramid.

Case 3. There is an *n*-gon (with $n \geq 4$) and at least two vertices, X and Y not in the plane of the *n*-gon, one on each side of it. Then the polyhedron is a bipyramid, and the segment $\overline{XY}$ is a diagonal.

Case 4. There is an *n*-gon (with $n \geq 4$) and at least two vertices not in the plane of the *n*-gon with both on the same side of it. In Figure 3.12, ABCD... is the *n*-gon, and E and F are the two vertices not in the plane of the *n*-gon. Since A, B, and C form an angle, we can't have both A, B, E, and F coplanar and B, C, E, and F coplanar. Without loss of generality, assume that B, C, E, and F are not coplanar. We use a proof by contradiction to show that A, B, E, and F must be coplanar. Suppose not. Consider the *half plane* P_1 starting from the line $\overleftrightarrow{AB}$ and going through E, the half plane P_2 starting from the line $\overleftrightarrow{AB}$ and going through F, and the half plane P_3 starting from the line $\overleftrightarrow{AB}$ and going through C and D. One of P_1 and P_2 is between the other one and P_3 since they are on the same side of P_3. Say P_2 is in between, as in Figure 3.13. Then the longer of the line segments $\overline{AF}$ or $\overline{BF}$ must be an interior diagonal. (In Figure 3.13, it is $\overline{AF}$.) Contradiction. Thus A, B, E, and F must be coplanar. The same argument applies to C, D, E, and F. If $n = 4$, the polyhedron is a modified triangular prism and we are done.

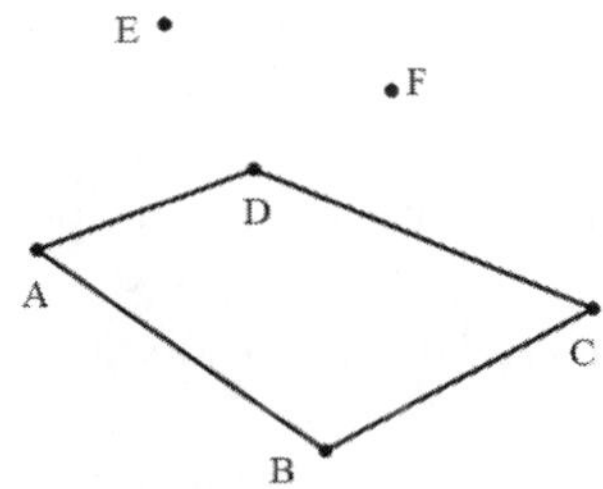

Figure 3.12. An *n*-gon and two vertices not in its plane.

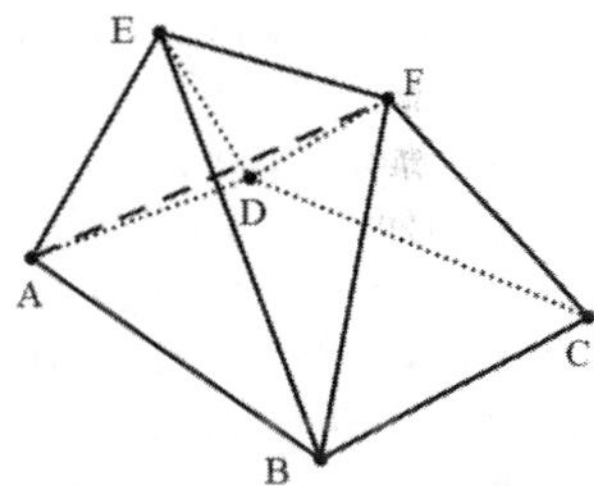

Figure 3.13. A, B, E, and F are not coplanar.

What happens if the polygon has more than four vertices, say A, B, C, D, and G? By the previous paragraph, E and F need to be coplanar with A and B, with C and D, and with D and G. But C, D, and G form an angle, making this impossible. So we have just a quadrilateral A, B, C, and D and so a modified triangular prism. □

Note: There are nonconvex polyhedra with no interior diagonals. We will discuss two in Section 4.2 when we answer Problems 3.4 (e) and 3.4 (i).

Problem 3.2. How can we count the number of interior diagonals of a convex polyhedron?

Problem 2.3. (repeated) For every number k between $n-3$ and $\frac{n(n-3)}{2}$, is there an n-gon with k diagonals?

Pick a vertex of a convex n-gon (V in Figure 3.14) and move it incrementally into the polygon cutting off diagonals one at a time. (In Figure 3.14, this will eliminate 8 of the 14 diagonals.) When you have cut off as many diagonals this way without cutting the polygon into two pieces, pick subsequent vertices and repeat until the minimum number of diagonals, $n - 3$, is reached.

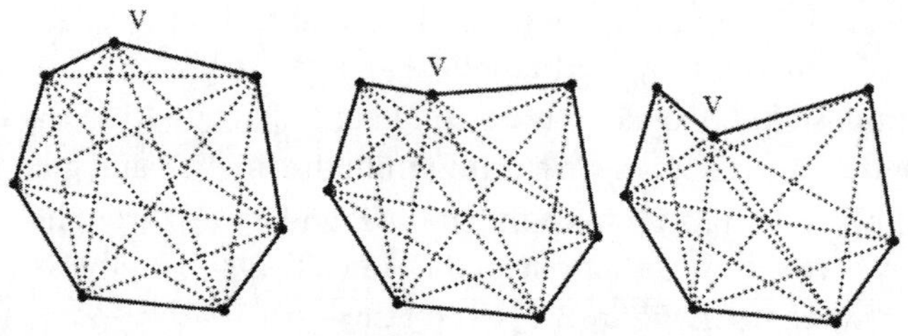

Figure 3.14. Eliminating diagonals in a polygon.

We know that a (nonconvex) polygon with n vertices has at least $n-3$ diagonals. We saw that with the exception of pyramids and modified triangular prisms, all convex polyhedra have interior diagonals. A triangular bipyramid has exactly one interior diagonal. (See Figure 2.15.) The next problem looks for polyhedra with a specified number of interior diagonals.

Problem 3.3. For each k construct a (possibly nonconvex) polyhedron with exactly k interior diagonals.

Problem 2.4 (a). (repeated) Explore whether polygons with n vertices and k interior points always have the same number of triangles in a triangulation. If so, find a formula for the number of triangles in a triangulation in terms of n and k.

All triangulations of polygons with interior points have $n + 2k - 2$ triangles, where n is the number of vertices of the polygon and k is the number of interior points. Some reasoning from an illustration may be more insightful than a formal proof using induction. In Figure 3.15, we modify the hexagon with three interior points to form a nonagon (shown with solid edges) with three exterior triangles to fill out the original hexagon. From Theorem 2.3, a nonagon has $9 - 2 = 7$ triangles. There are thus $10 = 7 + 3 = 6 + 2(3) - 2$ triangles in the completed triangulation. In general, we modify an n-gon with k interior points to have an $(n+k)$-gon with k exterior triangles to fill out the original n-gon. The $(n+k)$-gon has $n + k - 2$ triangles in its triangulation. When we add the k exterior triangles, we get $n + 2k - 2$ triangles.

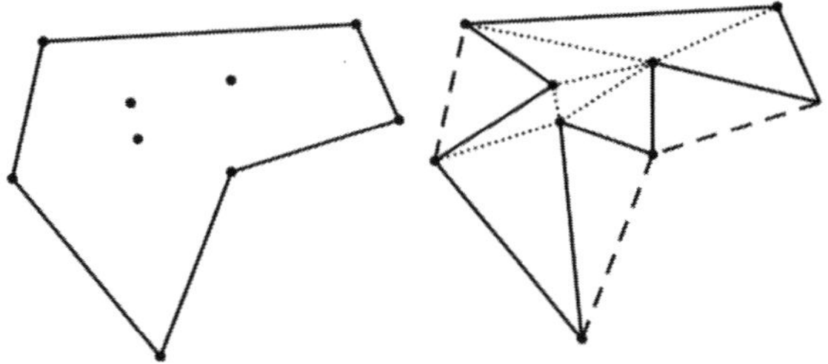

Figure 3.15. Modifying a polygon to include interior points.

You may have noticed a problem with the previous argument: What if there are so many interior points that we can't form exterior triangles? An induction proof based on the number of interior points can get around this problem. In the induction step we only need to build one exterior triangle to reduce the number of interior points by one.

Georg Alexander Pick (1859–1942) investigated polygons whose vertices are *lattice points*, that is, the coordinates of all the vertices are integers. He found a relationship between their areas and the number of interior lattice points and of lattice points that are vertices or on the edges of the polygons (*boundary points*). For instance, the polygon of Figure 3.16 has area 36.5 square units, 28 interior lattice points, and 19 boundary points.

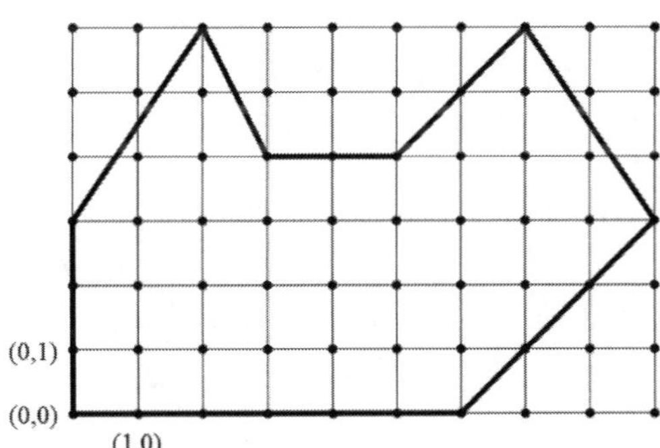

Figure 3.16. Polygon with lattice points for vertices.

Problem 3.4 (a). Draw a variety of polygons with lattice points as vertices and find their areas, the number of interior lattice points, and the number of boundary points. Is there a relation among these three numbers?

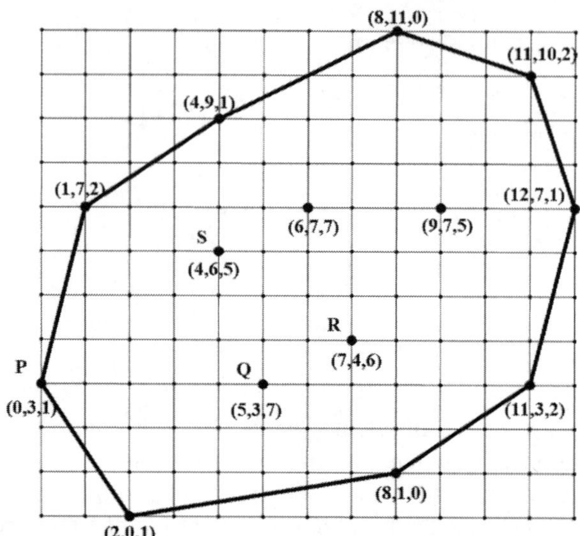

Figure 3.17. Fourteen points with indicated heights.

A more important application of triangulations of sets of points involves computer generated surfaces. Suppose a surveyor has measured the coordinates for a set of geographical places $P_1 = (x_1, y_1, z_1)$, $P_2 = (x_2, y_2, z_2)$, ..., $P_n = (x_n, y_n, z_n)$. Now we want to make a realistic computer representation of the surface. We can place the points on the xy-plane, triangulate them, and then raise the points to the appropriate heights. In the process, the triangles will be tilted and, we hope, represent the actual terrain. Figure 3.17 provides a simplified example with fourteen points. Even with only fourteen points there are thousands

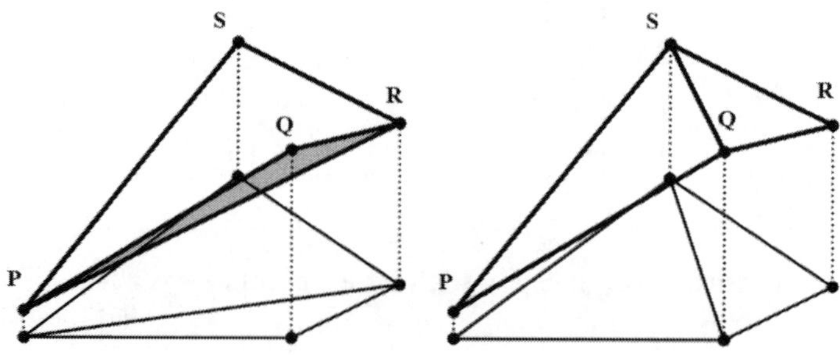

Figure 3.18. Two possible triangulations of P, Q, R, and S with heights shown.

of triangulations possible, but not all are likely representations of the terrain. Figure 3.18 shows the two possible triangulations of the quadrilateral PQRS together with the heights the points would have. The one on the left has a long narrow (partially shaded) valley from P to R, which is less likely than the rise from P up to the relatively level area of $\triangle$QRS.

Problem 3.4 (b). Explore geometric ways that a computer might use to decide which triangulation might be the most realistic for the fourteen points in Figure 3.17.

Another modification of the triangulation problem is for polygons with polygonal holes, as in Figure 3.19. This will correspond in Section 3.4 to art galleries and museums with some interior rooms without art, and so areas that don't need guards.

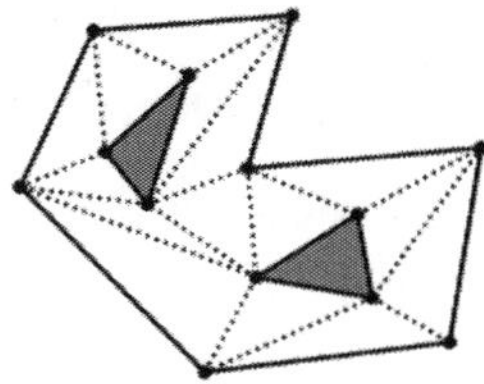

Figure 3.19. A triangulation of a polygon with two (shaded) holes.

Problem 3.4 (c). Explore whether polygons with n vertices (including the vertices of the holes) and h interior holes made of polygons always have the same number of triangles in a triangulation. If so, find a formula for the number of triangles in a triangulation in terms of n and h.

Problem 2.4 (b). (repeated) Find the number of ways to triangulate a convex hexagon and a convex heptagon. Describe a method (or find a formula) to find the number of ways to triangulate a convex n-gon.

Let T_n be the number of ways to triangulate a convex n-gon. We'll use the triangulation values we saw in introducing Problem 2.4 (b) in Chapter 2 of $T_3 = 1$, $T_4 = 2$, and $T_5 = 5$ to compute T_6 and with that, compute T_7. In each triangulation of the hexagon, any given edge, say $\overline{AB}$ in Figure 3.20, is a side of exactly one triangle. Let's investigate what options occur with each of the four choices for the third vertex of the triangle with A and B. For $\triangle$ABC, the remaining polygon is a pentagon, giving $T_5 = 5$ ways to continue a triangulation. The same happens with $\triangle$ABF. For $\triangle$ABD or $\triangle$ABE, the rest of the vertices are split into two parts, a triangle and a quadrilateral. So each of these options give $T_3 \cdot T_4 = 1 \cdot 2 = 2$ additional ways. In total, we have

$$T_6 = T_5 + T_3 \cdot T_4 + T_4 \cdot T_3 + T_5 = 5 + 2 + 2 + 5 = 14.$$

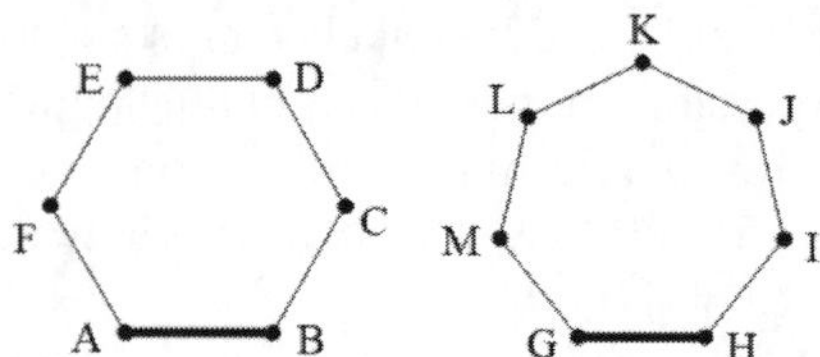

Figure 3.20. A hexagon and a heptagon.

Similar reasoning for the heptagon and side $\overline{GH}$ in Figure 3.20 gives

$$
\begin{aligned}
T_7 &= T_6 + T_3 \cdot T_5 + T_4 \cdot T_4 + T_5 \cdot T_3 + T_6 \\
&= 14 + 1 \cdot 5 + 2 \cdot 2 + 5 \cdot 1 + 14 \\
&= 42.
\end{aligned}
$$

This reasoning gives us a recursive, albeit cumbersome, way to determine a few higher values of T_n. For instance,

$$
\begin{aligned}
T_8 &= T_7 + T_3 \cdot T_6 + T_4 \cdot T_5 + T_5 \cdot T_4 + T_6 \cdot T_3 + T_7 \\
&= 42 + 1 \cdot 14 + 2 \cdot 5 + 5 \cdot 2 + 14 \cdot 1 + 42 \\
&= 132.
\end{aligned}
$$

Finding a closed formula for T_n is a more challenging problem with an interesting history. There are many situations counted by this sequence of numbers, called the Catalan numbers. They are named after the Belgian mathematician Eugene Charles Catalan (1814–1896), who studied these numbers and found and proved a closed formula for them. However, he wasn't the first to find this sequence or the formula. Leonhard Euler (1707–1783), the most prolific mathematician of all time, found the sequence and the formula for the number of triangulations. (Euler, however, didn't prove the formula. Gabriel Lamé (1795–1870) questioned Euler's answer and gave a proof.) And Euler wasn't alone at that time—the Mongolian mathematician Ming An-Tu (1692–1763) also found the number of ways to triangulate convex polygons. Mathematicians from different cultures often find the same problems challenging and provide independent solutions and proofs. While mathematics may not be a universal language, it can transcend languages and cultures in remarkable ways.

The proof of the formula for the Catalan numbers goes beyond the level of this book. The formula makes use of factorials, written $n!$, which we define recursively after giving some examples. After that, Theorem 3.2 gives the formula for the number of triangulations.

Examples. $3! = 3 \cdot 2 \cdot 1 = 6$. Similarly, $4! = 4 \cdot 3 \cdot 2 \cdot 1 = 24$ and $5! = 5 \cdot 4 \cdot 3 \cdot 2 \cdot 1 = 120$. By convention, $0! = 1$.

Definition. $0! = 1$, $(n + 1)! = (n + 1)(n!)$.

Theorem 3.2. The number of ways to triangulate a convex polygon with $n = k + 2$ vertices is $T_n = \frac{(2k)!}{k!(k+1)!}$. (This is the kth Catalan number, C_k.)

Examples. $T_8 = C_6 = \frac{12!}{6!7!} = \frac{479001600}{(720)(5040)} = 132$. $T_9 = C_7 = \frac{14!}{7!8!} = \frac{87178291200}{(5040)(40320)} = 429$.

Proof. See [11, 9–11].   □

The Catalan numbers count many different types of patterns. We'll give two further examples here—you can find over sixty other examples in [31].

We can count the number of paths going only right and up in an $n \times n$ lattice of points where the paths can't go above the diagonal. Figure 3.21 illustrates the five ways for a 4×4 array. Note that the first segment of the path from the lower left has to be to the right and the final segment must go up to the upper right. So with a 4×4 array, only the middle four segments have any choices.

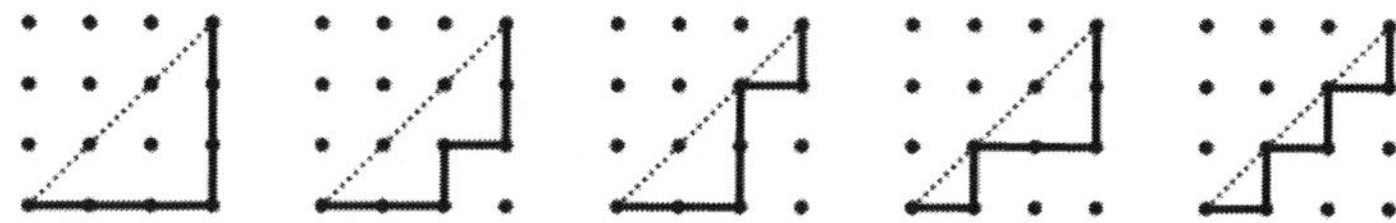

Figure 3.21. Possible right/up paths on a lattice of points.

We use parentheses to specify the order in which we do operations. For instance, $(1 - 2) - 3$ and $1 - (2 - 3)$ are different expressions with different values, -4 and 2, respectively. With two pairs of parentheses and four variables, we can have $((a - b) - c) - d$, $(a - (b - c)) - d$, $(a - b) - (c - d)$, $a - ((b - c) - d)$, and $a - (b - (c - d))$—five ways, the same Catalan number as the paths in Figure 3.21. In fact, there is a way to match each path with a different arrangement of parentheses. A segment moving right corresponds to a left parenthesis and an upward segment corresponds to a right parenthesis. To stay on or below the diagonal, at any point of the path, there have to have been at least as many rightward segments as upward ones. Similarly, to be a legitimate expression with parentheses, at any given point there have been at least as many left parentheses as right ones. Thus we can pair each path with a specific arrangement of parentheses, showing their total numbers are the same.

You can check that the next Catalan number, 14, counts the paths for a 5×5 array and the arrangements with three pairs of parentheses and five variables. Combinatorics frequently uses the reasoning we just did—matching the counting for two situations to show that they are equal.

Remark. Just because the ways we write the parentheses differ doesn't mean that the resulting numbers differ. For instance, for any numbers a, b, c, and d, $(a - (b - c)) - d = a - (b - (c - d))$. Although it is not a discrete geometry problem, you could investigate the maximum number of different numbers you can get using subtraction, parentheses, and n different numbers.

Now that we know the number of triangulations of a convex polygon, we can venture into triangulations of nonconvex polygons.

Problem 3.4 (d). What is the largest and smallest number of triangulations of a nonconvex polygon with n vertices?

Problem 2.4 (c). (repeated) Find a relationship between the vertices, edges, and faces of pyramids, prisms, bipyramids, antiprisms, and general convex polyhedra.

Tables 3.2, 3.3, 3.4, and 3.5 give, respectively, the number of vertices, edges, and faces for pyramids, prisms, bipyramids, and antiprisms having an n-gon as a base.

Table 3.2. Vertices, edges, and faces of pyramids.

n-gon base	3	4	5	n
vertices	4	5	6	$n + 1$
edges	6	8	10	$2n$
faces	4	5	6	$n + 1$

Table 3.3. Vertices, edges, and faces of prisms.

n-gon base	3	4	5	n
vertices	6	8	10	$2n$
edges	9	12	15	$3n$
faces	5	6	7	$n + 2$

Table 3.4. Vertices, edges, and faces of bipyramids.

n-gon base	3	4	5	n
vertices	5	6	7	$n + 2$
edges	9	12	15	$3n$
faces	6	8	10	$2n$

Table 3.5. Vertices, edges, and faces of antiprisms.

n-gon base	3	4	5	n
vertices	6	8	10	$2n$
edges	12	16	20	$4n$
faces	8	10	12	$2n + 2$

Tables 3.3 and 3.4 have a curious property: the number of vertices in one equals the number of faces in the other, and they agree on the number of edges. Is that a coincidence?

Problem 3.4 (e). Look for other pairs of polyhedra besides bipyramids and prisms that switch the number of vertices and faces. Given a polyhedron, can we find another polyhedron with this relationship?

The values in these tables may well suggest the general relationship that Leonhard Euler described in a letter in 1750. It took many years and the development of the field of topology to flesh out his argument to a full proof. We will state it for convex polyhedra and sketch a proof, although the formula applies more generally.

Theorem 3.3 (Euler's formula). Let V be the number of vertices, E the number of edges, and F the number of faces of a polyhedron. If the polyhedron is convex, then $V - E + F = 2$.

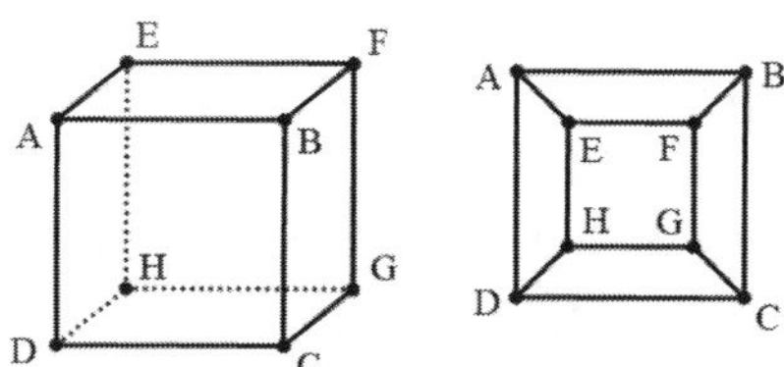

Figure 3.22. A cube and its flat projection.

Sketch of proof. We convert the edges and vertices of a convex polyhedron into what mathematicians call a *graph*—a flat network of vertices and edges where no edges intersect except at vertices. (This is a different use of the word "graph" from the familiar graphs of equations in algebra.) For example, Figure 3.22 converts the conventional picture of a cube on the left to the graph on the right. To do this we use a viewpoint from a point close to the center of a (transparent) face, but outside of the convex polyhedron. In the right of Figure 3.22, the vantage point is closer to your eye than the front of the cube, face ABCD. The faces of the graph are a distorted match of the faces of the polyhedron, except for the face closest to the point of view, which would cover all of the others. Alternatively, if we can count the region outside the graph as a face, the number of vertices, edges, and faces of the graph match those of the polyhedron.

Compute $V - E + F$ for the graph. Now we erase one edge at a time so as to keep the rest of the edges connected and retain at least one interior face. Sometimes the elimination of an edge turns two formerly adjacent faces into one face. (This happens after removing the first and second edges in Figure 3.23.) Then the new value of $V - E + F$ remains the same. The other option, as after the second edge is removed in Figure 3.23, is that an edge sticks out to a lone vertex. We agree to delete that vertex and edge as the next step, which also means that $V - E + F$ remains the same. Eventually we reduce the graph to a polygon. We know for any polygon the number of vertices equals the number of edges. By our counting, there will be two faces: the interior and the exterior. Thus $V - E + F = V - E + 2 = 2$. $\qquad\square$

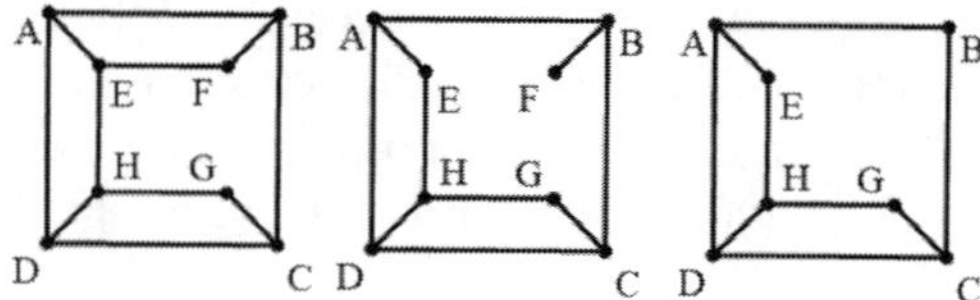

Figure 3.23. Removing edges of a projected polyhedron.

Armed with Euler's formula, if we know two of the three numbers V, E, and F, we can find the other one. Are there other relations among these numbers? What values of them are possible?

Problem 3.4 (f). Investigate equalities or inequalities linking V and E or E and F. Investigate what values of V, E, and F are possible.

The trick of turning the vertices and edges of the polyhedron into a two-dimensional graph depends on the convexity of the polyhedron so that the projected edges don't cross. Figure 3.24 shows some nonconvex polyhedra. The first two are *stellated polyhedra*; that is, we erect a pyramid on each face of the underlying polyhedron. In the second one, the pyramid pointing out toward the reader is shaded. The third shape has the peculiarity of having a hole. Naturally, we ask whether Euler's formula works for them, even if the sketch of an argument above doesn't work.

Figure 3.24. Two stellated polyhedra and a polyhedron with a hole.

Problem 3.4 (g). Investigate $V-E+F$ for nonconvex polyhedra to determine conditions needed for a polyhedron to satisfy Euler's formula. For those, if any, that don't satisfy Euler's formula, look for a more general formula.

Problem 2.4 (d). (repeated) Find a relationship between the number of vertices of a convex polyhedron and the sum of all the angles of all of its faces.

Let's start with two familiar families of polyhedra, pyramids and prisms. Recall from Chapter 2 after the statement of Problem 2.4 (b) that an n-gon has an angle sum of $180°(n-2)$. A pyramid has n triangles and an n-gon. However, we want the angle sum in terms of the number of vertices, $V = n + 1$. Its angle sum is $180°n + 180°(n - 2) = 360°(n - 1) = 360°(V - 2)$. A prism has n rectangles, each with $360°$ angle sum, two n-gons as bases, and $2n = V$ vertices. Its angle sum is $360°n + 2(180°)(n - 2) = 360°(2n - 2) = 360°(V - 2)$, the same value in terms of V. René Descartes found this formula more generally when he was in his twenties, but never published it. We can derive Descartes' formula from Euler's formula (which came over 100 years later), illustrating the beautiful connectivity of mathematical ideas.

Theorem 3.4 (Descartes' formula). A convex polyhedron with V vertices has an angle sum of $360°(V - 2)$.

Proof. We can rewrite Euler's formula $V - E + F = 2$ to get the value of Descartes' formula: $V - 2 = E - F$ and so $360°(V - 2) = 360°(E - F)$. However, it isn't so obvious what the right side has to do with the angle sum. The angle sum of each face depends on how many sides it has. Let F_3 be the number of triangular faces, F_4 the number of quadrilateral faces, etc. We know a polygon with i sides has angle sum $180°(i - 2)$. Now $F_3 + F_4 + \ldots + F_k = F$, where the face with the largest number of sides has k sides (and some of the terms could be 0). Triangular faces have three edges, quadrilateral faces have four edges, etc. At first glance, we might think the number of edges, E, would be $3F_3 + 4F_4 + \ldots + kF_k$, but each edge is a side of two faces. So $2E = 3F_3 + 4F_4 + \ldots + kF_k$. Then $E = \frac{3}{2}F_3 + \frac{4}{2}F_4 + \ldots + \frac{k}{2}F_k$. Now subtract F from the left side of this last equation and $F_3 + F_4 + \ldots + F_k$ from the right side. This gives us $E - F = \frac{1}{2}(F_3 + 2F_4 + \ldots + (k - 2)F_k)$. Multiply both sides by $360°$ to get $360°(E - F) = 180°(F_3 + 2F_4 + \ldots + (k - 2)F_k)$. Note the coefficient for each F_i is $i - 2$ and $180°(i - 2)$ is the angle sum of an i-gon. Thus the right side of the last equality is the angle sum of all the faces, showing Descartes' formula. $\square$

Problem 3.4 (h). How can we modify Descartes' formula for nonconvex polyhedra?

The last two problems of Section 2.2 considered splitting a polyhedron into tetrahedra. As often happens shifting from two dimensions to three dimensions, things get more complicated going from triangulations to tetrahedralizations. Part of the challenge is visualization, but in this case, some of the answers change dramatically.

Problem 2.4 (e). (repeated) Describe a way to tetrahedralize every pyramid. Do we always get the same number of tetrahedra?

For a pyramid, note that every tetrahedron needs to use the apex as one of its vertices. Then the other three vertices will be vertices of the base forming a triangle there. Those triangles triangulate the base, and we already know that every polygon can be triangulated. Even more, the number of triangles in such a triangulation is $n - 2$ when the polygon has n sides. So every tetrahedralization of a pyramid with an n-gon as a base has $n - 2$ tetrahedra.

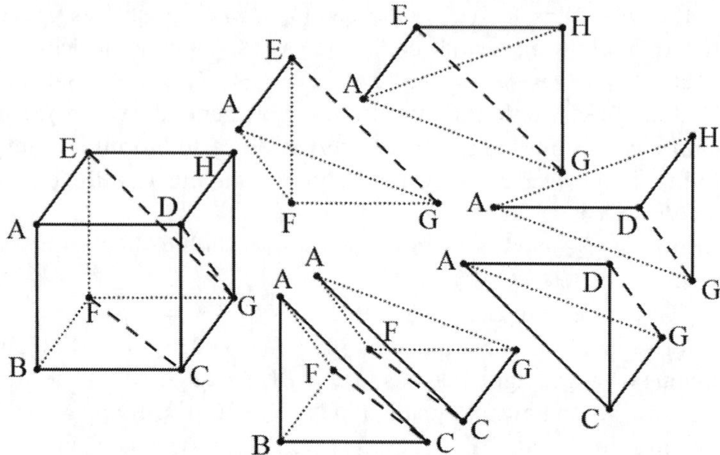

Figure 3.25. Common vertex tetrahedralization of a cube.

Problem 2.4 (f). (repeated) Can every convex polyhedron be tetrahedralized? If so, do we always obtain the same number of tetrahedra? If not, can we bound the minimum number of tetrahedra in terms of some aspect of the polyhedron?

Our success with pyramids suggest trying something similar for a convex polyhedron: pick a vertex, say A, and then triangulate all the faces **not** adjacent to A. Now construct tetrahedra with A as one vertex and the other three vertices forming one of those triangles. Figure 3.25 illustrates this process with a cube, where the vertex labelled A is the chosen vertex and the dashed edges $\overline{CF}$, $\overline{DG}$, and $\overline{EG}$ make the triangulations of the three squares not adjacent to A. This gives six tetrahedra, whose volumes together equal the volume of the cube. Call this way to tetrahedralize a polyhedron the *common vertex method.*

However, not all tetrahedralizations of a polyhedron have the same number of tetrahedra. For instance, if we use the inscribed tetrahedron ACFH, as in Figure 3.26, we can split the cube into five tetrahedra instead of six.

Even the common vertex method can yield different numbers of tetrahedra when vertices have different properties. All bipyramids have two different tetrahedralizations and most have a version (or versions) mixing these two. In one version of the common vertex method, all the tetrahedra share as an edge the interior diagonal between the two apices (plural of apex) of the pyramids. (We can chose either apex as the common vertex.)

Another type tetrahedralizes each of the pyramids separately, which includes the common vertex method where the common vertex is not an apex. (See Figure 3.27.) For a bipyramid with an n-gon as the base, the first method uses n tetrahedra, while the second method uses $2(n - 2)$ tetrahedra. In the bipyramid on the

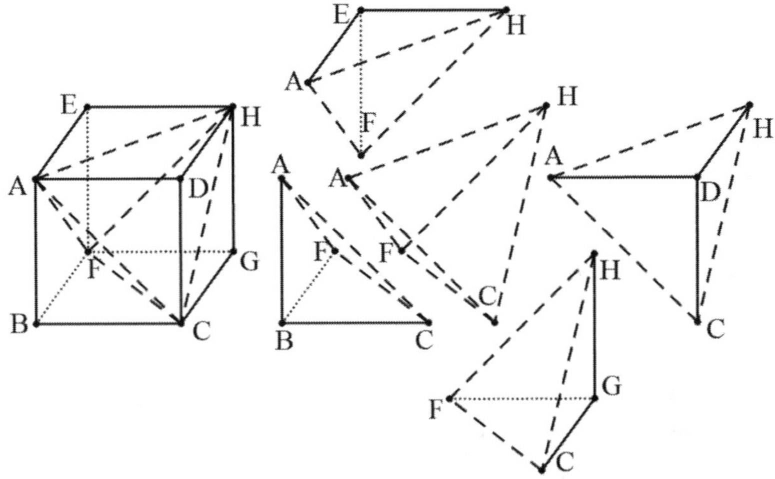

Figure 3.26. Another tetrahedralization of a cube.

left of Figure 3.27, the interior diagonal between the apices is shown as a dashed line. The middle three tetrahedra illustrate the first method of tetrahedralization and the two tetrahedra at the right illustrate the second method. Except for the case $n = 3$, the first method gives the minimum bound for bipyramids. The second method may seem a good candidate to give a maximum bound, but we can tweak it a bit.

For a bipyramid with a base having at least six sides, select three nonadjacent vertices on the base and use the first method to tetrahedralize the triangular

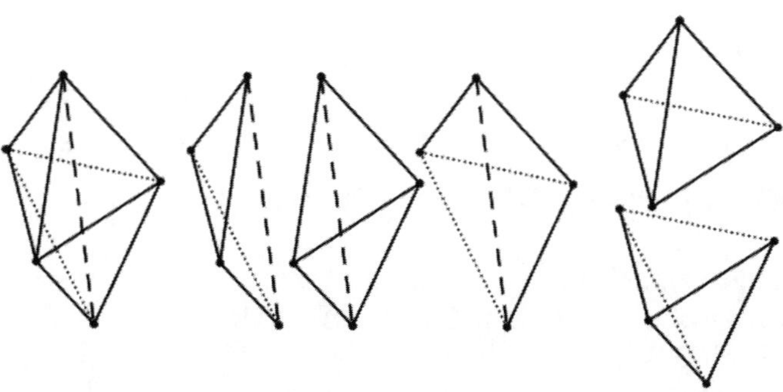

Figure 3.27. Two ways to tetrahedralize a bipyramid.

bipyramid they form with the two apices. Then use the second method to tetrahedralize the rest. That gives $2n - 1$ tetrahedra, which appears to be a maximum.

For a tetrahedralization of any polyhedron, the exterior triangles of the tetrahedra need to cover the surface of the polyhedron. The surface of the polyhedron will thus be triangulated by the exterior triangles of the tetrahedra. This gives a minimum bound on the number of tetrahedra. With the exception of the polyhedron being a tetrahedron, no tetrahedron in the tetrahedralization can have all four of its triangles on the surface of the polyhedron. So, the minimum number of tetrahedra for other polyhedra is $\frac{1}{3}$ the number of triangles in triangulations of all of its faces. This minimum can happen for a triangular bipyramid. This lower bound doesn't generally work. The six faces of a cube split into twelve triangles. Thus any tetrahedralization has to have at least $4 = \frac{1}{3}(12)$ tetrahedra. In actuality, the minimum number of tetrahedra for a cube is five. When that happens, four of the tetrahedra have three triangles on the surface of the cube, covering the entire surface. The fifth tetrahedron has no surface triangles, as in Figure 3.26.

Problem 3.4 (i). Investigate tetrahedralizing nonconvex polyhedra.

3.3 Distances and Points

Problem 2.5 (a). (repeated) For some n is there an arrangement of n points in the plane with fewer distances than $\left\lfloor \frac{n}{2} \right\rfloor$, the number for a regular n-gon?

The symmetry of a regular polygon leads to many pairs of points with the same distance, whereas for the maximum number of distances in Problem 1.5 (a) we had no repeated distances. In addition to the square and regular heptagon that give the minimum number of distances, Figure 3.28 provides other arrangements for $n = 4$ and $n = 7$.

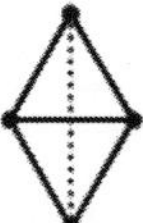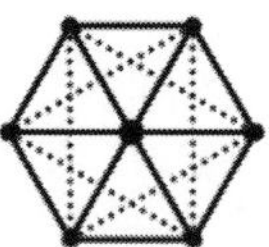

Figure 3.28. Alternative arrangements of four and seven points with two and three distances, respectively.

To reduce even further, we need more repeated distances. A first idea might be a square array of n^2 points, as in Figure 3.29. If the distance between adjacent horizontal points is 1, there are $(n - 1)^2$ horizontal pairs of points a distance of 1 apart, plus that many vertical pairs. We also get many horizontal and vertical pairs a distance of 2 apart, many diagonal pairs a distance of $\sqrt{2}$ apart, etc. But do we get fewer total distances? In Figure 3.29, all possible distances can be realized

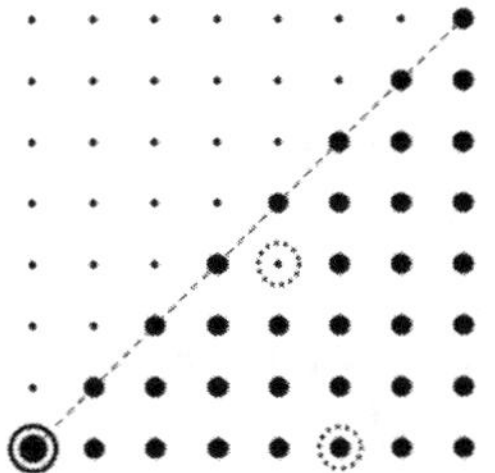

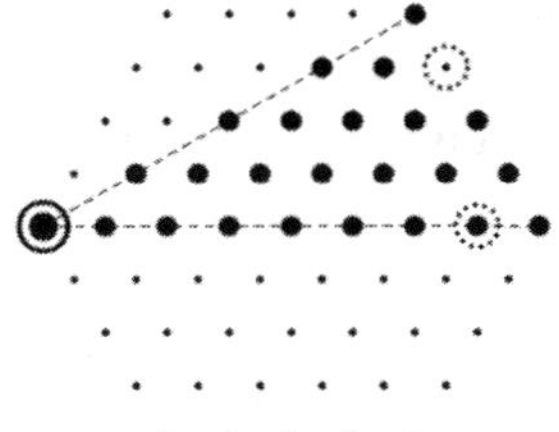

Figure 3.29. Distinct distances in a square arrangement of points.

Figure 3.30. Distinct distances in a hexagonal arrangement of points.

using the circled point in the lower left as one of the two points. Further by symmetry, the second point can be one of the large points on or below the dashed diagonal line. Unfortunately, because of the multiple points on the diagonal, that includes noticeably more than half the points. (This is reduced a little bit because of the Pythagorean theorem. For instance, the points with the dotted circles are the same distance from the lower left point since $4^2 + 3^2 = 5^2$.) For the 64 points in Figure 3.29, there are 34 distances, more than the 32 distances in a regular 64-gon.

A hexagonal arrangement of points, as in Figure 3.30, can reduce the number of distances below half the number of points. As in Figure 3.30, all the distances can be realized using the circled point on the far left together with one of the large points. The overall symmetry of the design allows us to eliminate the points below the horizontal dashed line. In addition, a more local symmetry allows us to eliminate the points above the diagonal dashed line. Finally, the law of cosines, a generalization of the Pythagorean theorem illustrated in Figure 3.31, can reduce the number of distances a bit further. The small dot with the dashed circle can be reached from the far left point by going five horizontal units, followed by three

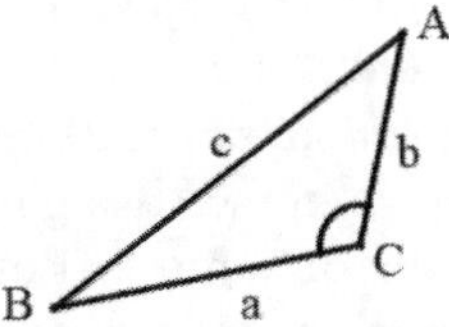

Figure 3.31. The law of cosines. $c^2 = a^2 + b^2 - 2ab\cos(C)$.

units at an angle. Using the law of cosines (Theorem 3.5 below), we get $5^2 + 3^2 - 2(5)(3)\cos(120°) = 25 + 9 + 15 = 49 = 7^2$. Thus, for these 61 points there are 23 distances, fewer than the 30 distances in a regular 61-gon.

Theorem 3.5 (Law of cosines). (Euclid, 300 BCE) In any (Euclidean) triangle $\triangle ABC$, $c^2 = a^2 + b^2 - 2ab\cos(C)$.

Proof. See Euclid *Elements*, Book II, Propositions 12 and 13 [14, 405–406] or, to use cosines, consult any recent textbook on trigonometry. $\square$

The Pythagorean theorem was proven approximately 2,500 years ago. Perhaps surprisingly, the law of cosines without trigonometric language was proven 2,300 years ago in what is the most influential mathematics book of all time, Euclid's *Elements*.

Problem 3.5 (a). Find a formula for the number of points in a hexagonal arrangement of points with k circular layers. (Figure 3.30 has four circular layers of points. With one layer of points there are seven points, the center one and six surrounding points.) Find a formula for the number of distances determined by k circular layers, ignoring the duplicates using the law of cosines.

Problem 2.5 (b). (repeated) Find arrangements of points in three dimensions with fewer distances than can be done in two dimensions with the same number of points.

Table 3.6. Minimum distances in two and three dimensions.

n (points)	4	5	6	7	8	12	20
2-d distances	2	2	3	3	4	6	10
3-d distances	1	2	2	3	3	3	5

The four vertices of a regular tetrahedron are all the same distance from each other, something that can't happen in the plane. Table 3.6 gives the minimum number or presumed minimum of distances in two and three dimensions for selected choices of n points. (As recently as 2011, the minimum values in two dimensions were proven just up to 14 points. The minimum number of distances in three-dimensional are even harder.) The minimum number of distances in three dimensions for 4, 6, 8, 12, and 20 points are for the vertices of the five regular polyhedra, which have many symmetries that reduce the number of different distances. (See Figure 3.32, shown in increasing number of faces.) If there is an arrangement of n points with a certain number of distances, say 20 points with 5 distances, than there is an arrangement of k points with $k < n$ having at most that many distances. But we might be able to get more points without needing more distances.

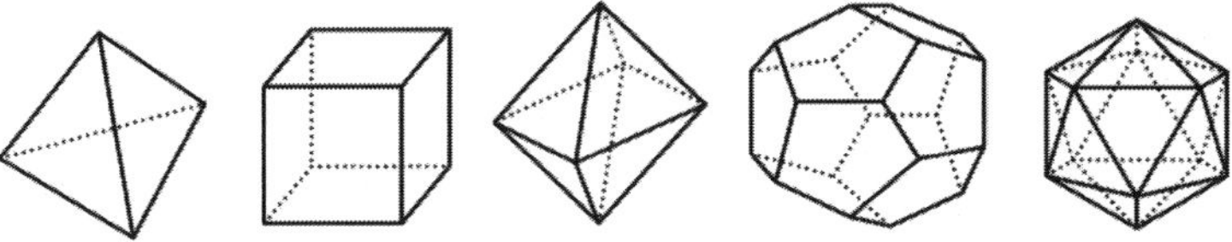

Figure 3.32. The five regular polyhedra: tetrahedron, cube, octahedron, dodecahedron, and icosahedron.

The preceding discussion, Table 3.6, and Problem 3.5 (a) raise a more general and much more difficult question about points and distances in three dimensions.

Problem 3.5 (b). Find an arrangement in three dimensions of more than twenty points that you think has a minimum number of nonzero distances. Try to describe a general arrangement in three dimensions to minimize the number of distances.

Problem 2.5 (c). (repeated) Find the minimum number of nonzero distances of n points in the plane using taxicab distance, for $n = 5, 6, 7,$ and 8.

The answer for all of these values of n in taxicab geometry is two. Figure 3.33 illustrates that we can have up to nine points in the plane having only two distances using the taxicab metric.

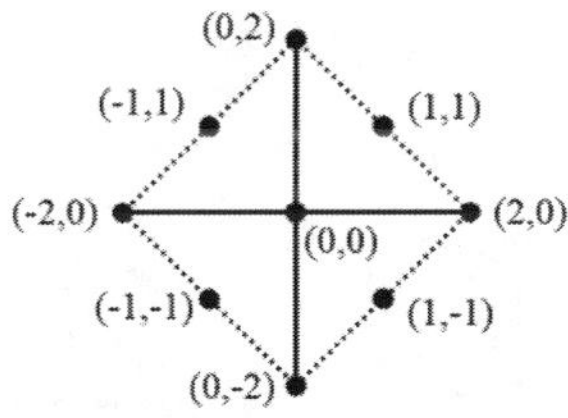

Figure 3.33. Nine points with only the nonzero taxicab distances of 2 and 4.

Problem 2.5 (d). (repeated) Make a conjecture about the maximum number of points in the plane that have a total of 3, 4, or 5 taxicab distances.

Figures 3.33 and 3.34 suggest that the maximum number of points for k nonzero taxicab distances is $(k + 1)^2$. In fact, this was proven in 2020 by undergraduate mathematics students.

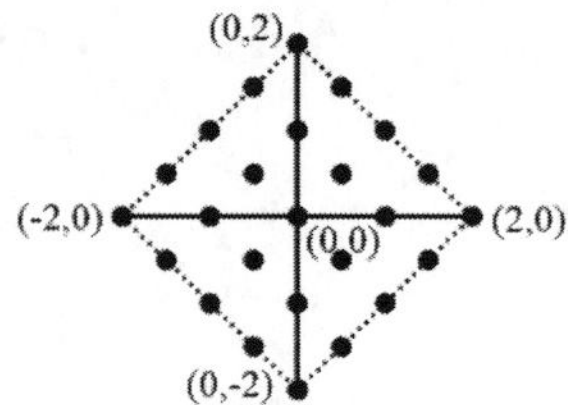

Figure 3.34. 25 points in the plane with four distances using the taxicab metric.

Theorem 3.6. (Balaji, Edwards, Loftin, Mcharo, Phillips, Rice, and Tsegaye, 2020) The maximum number of points in the plane determining k nonzero taxicab distances is $(k + 1)^2$.

Proof. See [3]. $\qquad\qquad\qquad\qquad\qquad\qquad\qquad\qquad\qquad\qquad$ □

Problem 2.5 (e). (repeated) Generalize taxicab distance to three dimensions and repeat Problem 2.5 (d) for three dimensions.

Definition. The *taxicab distance* between points (p, q, r) and (s, t, u) in three dimensions is $d_t((p, q, r), (s.t.u)) = |s - p| + |t - q| + |u - r|$.

The six vertices of a regular octahedron are all the same distance from each other, illustrated in Figure 3.35, so we can have up to six points with one nonzero distance. Figure 3.36 mimics Figure 3.33 in three dimensions to get 19 points having just two distances. Figures 3.37 places 44 points with just three distances.

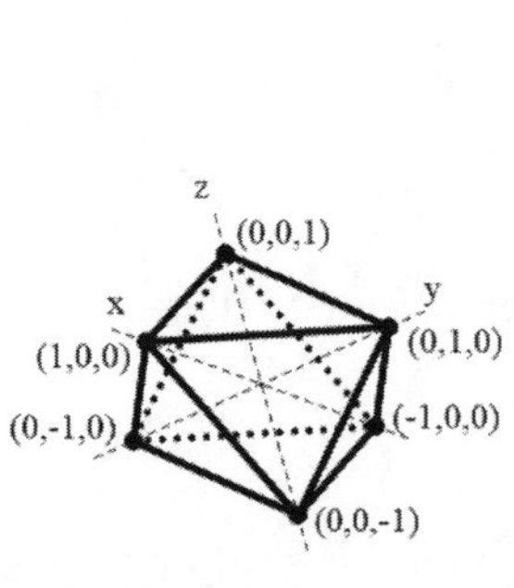

Figure 3.35. Octahedron.

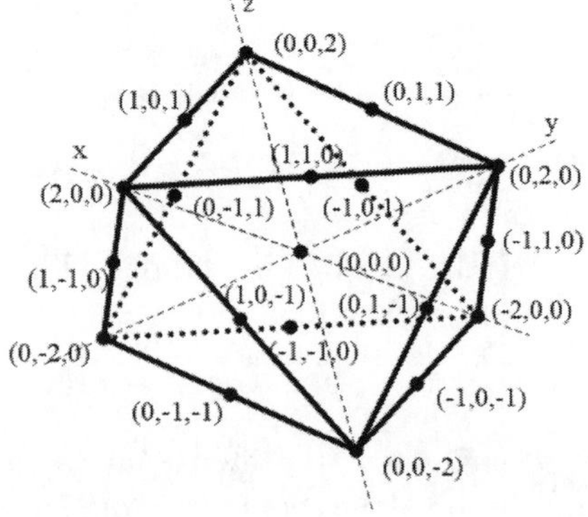

Figure 3.36. 19 points with two nonzero taxicab distances.

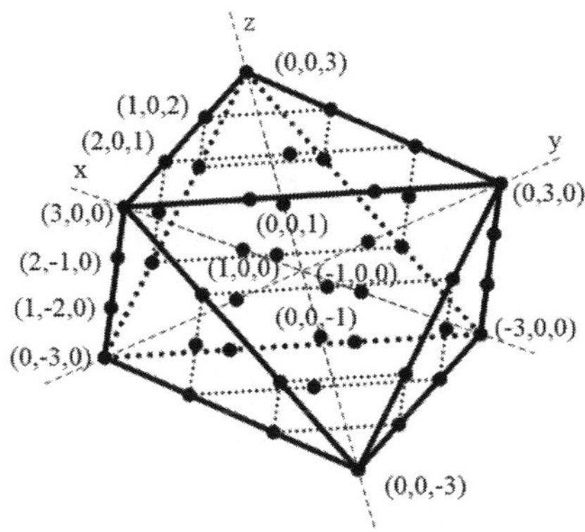

Figure 3.37. 44 points with 3 taxicab distances and with layers indicated.

We can use integer coordinates for all the points in Figures 3.35 to 3.37. For the octahedron, the points have the form (x, y, z), where $|x| + |y| + |z| = 1$ and x, y, and z are integers. With more distances we allow two or more possible values for the sum $|x| + |y| + |z|$, 0 or 2 for Figure 3.36 and 1 or 3 for Figure 3.37. Taking horizontal slices of each set of points (as indicated in Figure 3.37) we have $1 + 4 + 1 = 6$ for two distances, $1 + 4 + 9 + 4 + 1 = 19$ for three distances, $1 + 4 + 9 + 16 + 9 + 4 + 1 = 44$ for four distances, and for five distances $1 + 4 + 9 + 16 + 25 + 16 + 9 + 4 + 1 = 85$. Let's look for a formula.

Table 3.7 gives the differences and the differences of the differences, which suggests that the formula will be third degree. For k distances there are $\frac{2}{3}k^3 + 2k^2 + \frac{7}{3}k + 1 = \frac{2}{3}(k + 1)^3 + \frac{1}{3}(k + 1)$ points in this arrangement. (The $k + 1$ comes from including the zero distance.) The paper [3] conjectures that this is the correct value and this arrangement of points is the correct one. Even more, it gives a general formula for any number of dimensions and distances in the taxicab metric, suggesting Problem 3.5 (d).

Table 3.7. Maximum number of points in three dimensions with taxicab metric and differences.

distances	0	1	2	3	4
points	1	6	19	44	85
1st differ	-	5	13	25	41
2nd differ	-	-	8	12	16

Problem 3.5 (c) Generalize taxicab distance to four dimensions and repeat Problem 2.5 (d) for four dimensions.

Problem 2.5 (f). (repeated) Find an arrangement of unit circles that you think is packed as tightly as possible. Define some way to measure how tightly packed your arrangement is. Find another packing of unit circles that has a different measure than the first one, yet doesn't have any "holes" big enough to allow any more unit circles in the region.

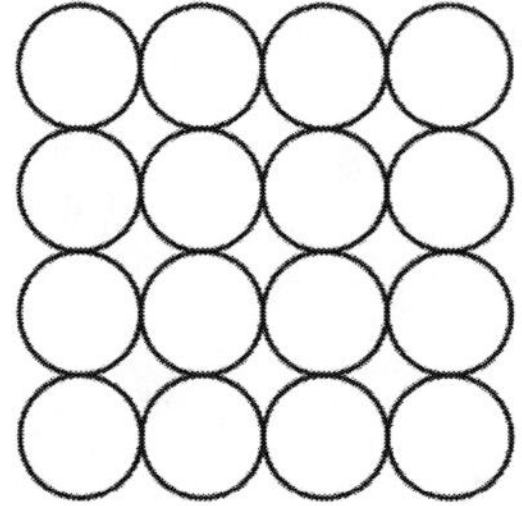

Figure 3.38. Square array of circles.

Figure 3.39. Hexagonal array of circles.

The "square" array of circles in Figure 3.38 may seem at first like a good choice for a tight packing of unit circles. However, the "hexagonal" array in Figure 3.39 appears to be a tighter packing. (It is no accident that we used a hexagonal arrangement of points in answering Problem 2.5 (a) and a hexagonal arrangement of circles to pack circles.) To verify the hexagonal packing is tighter we need to define some measure to move beyond appearances.

The intuition of the density of a packing is the percentage of the area covered by the circles. To make that intuition more precise, our definition considers polygons, as in Figure 3.40, whose vertices are the centers of the circles surrounding a given circle. Because the circle packings in Figure 3.40 are periodic, the overall density of the circle packing is the same as the density within the polygons. (Periodic means that the pattern repeats in a regular fashion in all directions.) For nonperiodic packing, the definition of overall density requires the concept of a limit from calculus.

Definition. The *density* of a circle packing in a given region is the ratio $\frac{\text{area of circles}}{\text{area of region}}$.

The area of the circles in the square in the left of Figure 3.40 is the equivalent of four unit circles, 4π. The square has area 4^2, so the density is $4\pi/16 \approx 0.7854$. The hexagonal region on the right of Figure 3.40 has three circles' worth of area, 3π. The hexagon is made of six equilateral triangles, each with a side of 2. Each triangle has a height of $\sqrt{3}$ and base of 2, and so an area of $\sqrt{3}$. Its density is

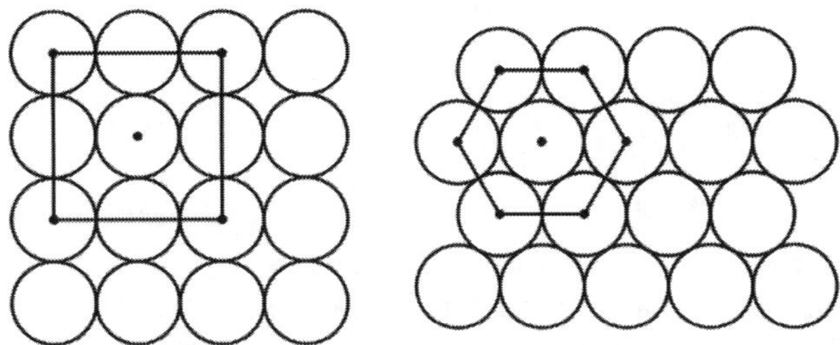

Figure 3.40. Polygons surrounding a circle in the arrays of Figures 3.38 and 3.39.

$3\pi/(6\sqrt{3}) \approx 0.9069$, noticeably higher than for the square packing. In 1773, Joseph-Louis Lagrange (1736–1813) proved that the hexagonal arrangement was the most dense among periodic circle packings that fill the plane. It took until 1942 for a complete proof by László Fejes Tóth (1915–2005) that this was the densest packing, whether the circles were periodic or not.

Instead of filling the entire plane, we can try packing unit circles in squares. Figure 3.41 shows the smallest squares that allow one, two and four nonoverlapping unit circles. The densities for the squares with one and four circles are the same as for the square periodic packing, 0.7854. The side of the square with two circles is $2 + \sqrt{2}$ and the density is $\dfrac{2\pi}{(2+\sqrt{2})^2} \approx 0.5390$.

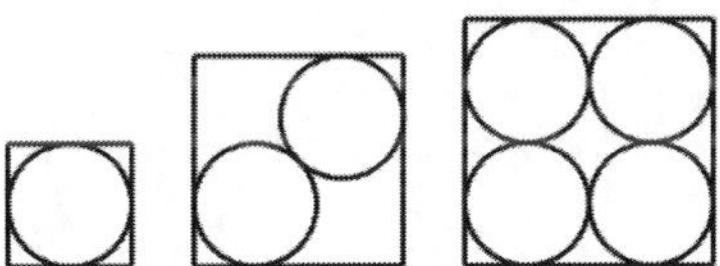

Figure 3.41. Smallest squares with one, two, and four unit circles.

Problem 3.5 (d). Find the dimensions of the smallest squares that contain n nonoverlapping unit circles, for $n = 3$, 5, 8, and 9. Find the corresponding density. (The densest packings for these and some other values of n have helpful symmetry.)

Problem 3.5 (e). Find some periodic packings of spheres and determine their densities.

Problem 3.5 (f). Define a taxicab circle and sphere. Explore circle and sphere packing in taxicab geometry.

3.4 The Art Gallery Problem

Problem 2.6 (a). (repeated) Find the maximum number of points needed to form a set of fortress points for a convex polygon with n vertices. Generalize to nonconvex polygons.

The solution of the art gallery problem put the guards at some vertices of the polygon. We can start with them for fortress points. Two examples may convince you of the general situation for convex polygons. For the pentagon in Figure 3.42, guard points at A and B can see everything outside of the "fortress" except the region in the triangle with dotted boundaries. To cover that area, we need a third guard somewhere on the side where point C is. For the hexagon in Figure 3.42, the guard points D, E, and F suffice. The number of fortress guards for convex polygons appears to be half of the vertices or a bit more. We denote this rounding up to the next integer with the *ceiling function*: $\lceil x \rceil$ is the least integer greater than or equal to x. For example, $\lceil \pi \rceil = 4 = \lceil 4 \rceil$. That is, we expect the number of fortress points needed in a convex n-gon is $\left\lceil \frac{n}{2} \right\rceil$.

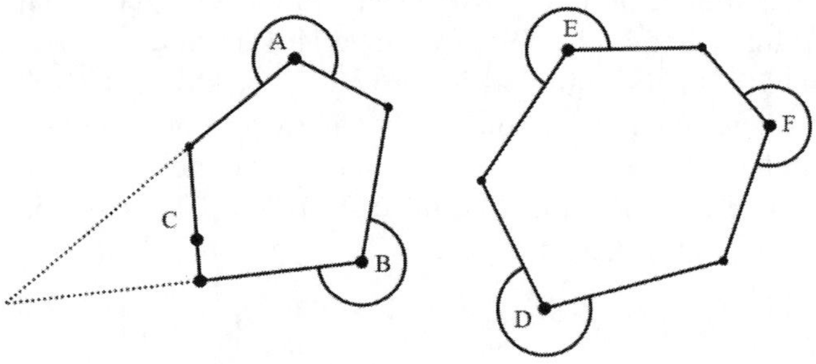

Figure 3.42. Guard points for a convex pentagon and hexagon.

Theorem 3.7. A convex polygon with n vertices needs $\left\lceil \frac{n}{2} \right\rceil$ fortress points placed at vertices so that no two fortress points are adjacent or just one pair is adjacent when n is odd.

Proof. Left to you.   □

Nonconvex polygons seem more interesting. The polygon in Figure 3.43 has thirteen vertices, but needs only four fortress points. It suggests that convex polygons require the most fortress points. That is exactly the opposite from guard points, where a convex polygon needed only one guard point.

Guarding a nonconvex part outside a fortress is like guarding the inside of an art gallery. Figure 3.44 adds segments in each of the indentations of the polygon of Figure 3.43. The polygon AEBFD is convex and so needs just the fortress points A, B, and D to cover all the region outside of that polygon. (It is called

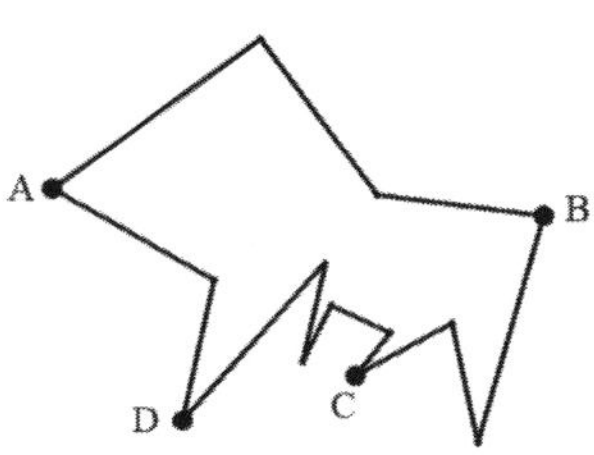

Figure 3.43. A nonconvex fortress.

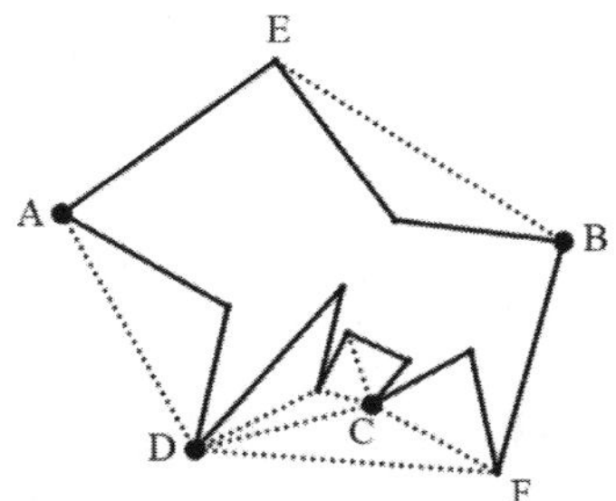

Figure 3.44. Figure 3.43 with segments triangulating its indented exterior.

the *convex hull* of the original polygon.) For the lower exterior part of the polygon, Figure 3.44 gives a triangulation. We can see that each triangle there has C or D (or both) as vertices. Thus C and D suffice for that indented part. Two mathematicians generalized this example into a general proof for Theorem 3.8. Instead of the technical proof, we give a sketch of why it is reasonable to think the exterior of a nonconvex polygon is not harder to guard than that of a convex polygon. (See [24, 146] for the proof.)

Theorem 3.8. (O'Rourke and Wood, 1983) A polygon with n vertices needs at most $\left\lceil \frac{n}{2} \right\rceil$ fortress points.

Idea for proof. Let P be a polygon with n vertices and Q be its convex hull. That is, the smallest convex polygon containing P. The vertices of Q are a subset of the vertices of P, say Q_1, Q_2, ..., Q_k, where $k \leq n$. By Theorem 3.7 we need either $\frac{k}{2}$ or $\frac{k+1}{2}$ of the vertices Q_i as fortress points for the region outside of Q. Each indentation of the exterior of P is bounded by some segment $\overline{Q_i Q_{i+1}}$ of Q. By Theorem 2.5, we need at most $\frac{1}{3}$ of the vertices of P between Q_i and Q_{i+1} to be fortress points and at least one of Q_i and Q_{i+1} are already fortress points for the region outside of Q. That is, in addition to the fortress points needed for Q, we need to add approximately $\frac{n-k}{3}$ more fortress points for P. (Some of the points Q_i are double counted since they are on the convex hull and on indentations. The technical proof avoids this problem.) For k even, this is at most $\frac{k}{2} + \frac{n-k}{3} = \frac{2n+k}{6} \leq \frac{3n}{6} = \frac{n}{2}$ fortress points. When k is odd, we have at most $\frac{k+1}{2} + \frac{n-k}{3} = \frac{2n+k+1}{6} \leq \frac{3n+1}{6} = \frac{n}{2} + \frac{1}{6}$ fortress points. However, the number of fortress points is an integer, so we can use the greatest integer function $\lfloor\ \rfloor$. Regardless of whether n is even or odd, $\left\lfloor \frac{n}{2} + \frac{1}{6} \right\rfloor = \left\lfloor \frac{n}{2} \right\rfloor \leq \frac{n}{2}$. $\qquad\square$

Problem 2.6 (b). (repeated) Define what a set of prison guard points is. Find the maximum number of points needed to form a set of prison guard points for a convex polygon with n vertices. Generalize to nonconvex polygons.

A prison guard standing on the wall of a prison can look inside and outside, but not through other walls of the prison. (For ease, we assume that the prison has just the outside wall and not, like real prisons, multiple inner walls.) This explanation suggests the following definition.

Definition. For a polygon P, a set of *prison guards* $\{P_1, P_2, \ldots, P_n\}$ is a set of points on the boundary of P (that is, vertices and points on the edges of P) so that for all points X of the plane, there is some P_i so that the only point of $\overline{P_iX}$ on the boundary of P is P_i or $\overline{P_iX}$ is part of the boundary of the polygon.

We need at least as many prison guards as the larger of the fortress guards and the art gallery guards. For convex sets, we know that will be the fortress guards, $\frac{n}{2}$ or $\frac{n+1}{2}$, depending on whether n is even or odd. Further, all fortress guards and art gallery guards can be placed at vertices of the polygon. We might immediately think that whichever set is bigger, the museum guards or the fortress guards, would therefore be a set of prison guards. However, as Figure 3.45 indicates, we may need more points than that. On the left of Figure 3.45, $\{B, X, Y\}$ is a set of guard points. In the middle, we see that $\{A, B, C, D\}$ is a set of fortress points. But neither gives a set of prison guards. A smallest set of prison guards is $\{A, B, C, D, Z\}$. A solution that will work for all polygons is to take the union of a set of guard points and a set of fortress points. In Figure 3.45, that set would be $\{A, B, C, D, X, Y\}$, a larger set than the minimal one. Problem 3.6 (a) asks you to find an example where the union is the smallest possible, so that Theorem 3.9 is the best we can expect in general.

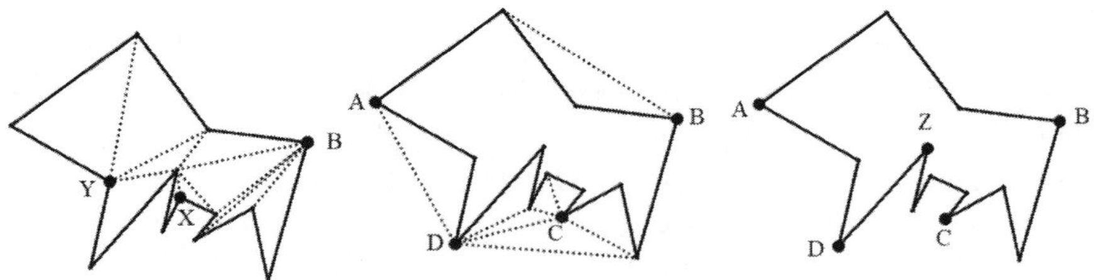

Figure 3.45. Minimal sets of guard points, of fortress points, and of prison guards.

Theorem 3.9. For any polygon, the union of a set of guard points and a set of fortress points is a set of prison guards.

Proof. Left for you.   □

Problem 3.6 (a). Find a polygon whose smallest set of prison guards is the union of a smallest set of guard points and a smallest set of fortress points, but no set of guard points and no set of fortress points forms a set of prison points.

Real buildings tend to have walls set at right angles to one another. Does that allow the number of guard points, fortress points, or prison guard points to be smaller for a general n-gon?

Problem 3.6 (b). Reconsider Problems 1.6, 2.6 (a), and 2.6 (b) where all angles of the polygon are right angles.

Problem 3.6 (c). Consider fortress points for convex polyhedra.

Problem 3.6 (d). Consider the art gallery problem for polygons with polygonal holes.

Problem 3.6 (e). Consider the art gallery problem for nonconvex polyhedra in three dimensions.

3.5 Geometric Patterns

Problem 2.7 (a). (repeated) Draw a number of maps and color them with the minimum number of colors so that no adjacent regions are the same color. Find conditions when the map requires at least four colors and when it can be colored with just two colors.

The left two maps in Figure 3.46 require four colors, while the right one only needs three colors. Each of these maps has a central region surrounded by regions. These suggest one condition for when a map requires four colors. We say a sequence of regions $R_1, R_2, \ldots, R_n$ are *cyclically adjacent* provided for $1 \leq i < n$, R_i and R_{i+1} are adjacent, as are R_n and R_1. In each of the maps in Figure 3.46, the ring of surrounding regions is cyclically adjacent.

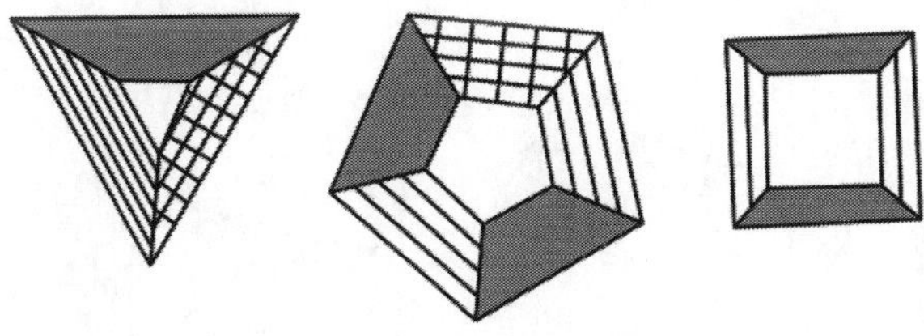

Figure 3.46. Two maps requiring four colors and one needing just three colors.

Theorem 3.10. A map with a region adjacent to an odd number of regions greater than two that are cyclically adjacent requires at least four colors.

Proof. Suppose region S is adjacent to each of the cyclically adjacent regions $R_1, R_2, \ldots, R_{2n+1}$, where $n \geq 1$. Since each R_i is adjacent to S, they must have different colors than the color for S. So, we need at least three colors for the mutually adjacent regions S, R_i, and R_{i+1} for $1 \le i < n$. To show we need four colors, suppose for a moment that three colors sufficed, say color a for S, b for R_1, and c for R_2. Then R_3 would have color b and so on, with the colors b and c alternating. That is, the regions with odd subscripts would have color b and those with even subscripts would have color c. But then R_{2n+1} would have color b, the same as R_1 to which it is adjacent, a contradiction. Thus we need to have at least four colors. $\qquad\square$

Theorem 3.11. If all boundaries of regions in a map are determined by straight lines that go from one side of the map to another, then the map needs only two colors.

Proof. (by induction on the number of lines) For the base case, suppose a map has just one line going from one side of the map to the other. There are then just two regions and so two colors suffice. For the induction step, suppose that any map with at most n lines determining the regions needs at most two colors. For a map with $n + 1$ lines, select one of these, call it k, (the dotted one on both sides of Figure 3.47). For the moment, ignore line k. By the induction hypothesis, we can use two colors to color the regions determined by the other n lines. (See the left side of Figure 3.47.) Now on one side of line k, keep the colors of regions the same, and on the other side of k, switch all the colors. (See the right side of Figure 3.47.) Each side can be colored with two colors. Further, adjacent regions separated by line k are now different colors. We thus have a two-coloring of the map with all $n + 1$ lines. By induction, we can do this with any number of lines. $\qquad\square$

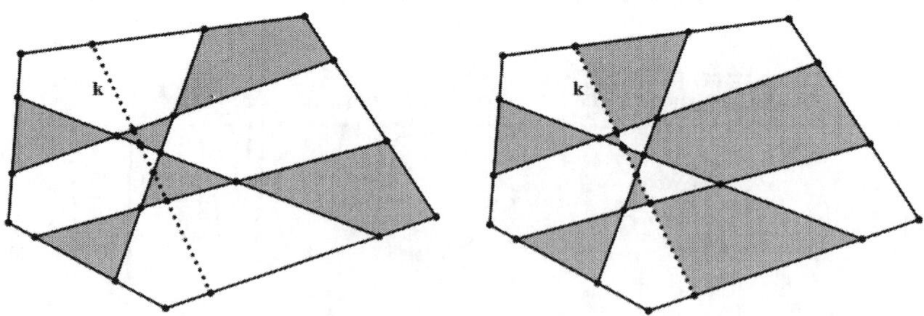

Figure 3.47. Using two colors to color a map with $n + 1$ lines.

The reasoning of Theorem 3.11 can be extended, as Figure 3.48 suggests. The key is that each intersection has an even number of regions surrounding it so we can alternate colors.

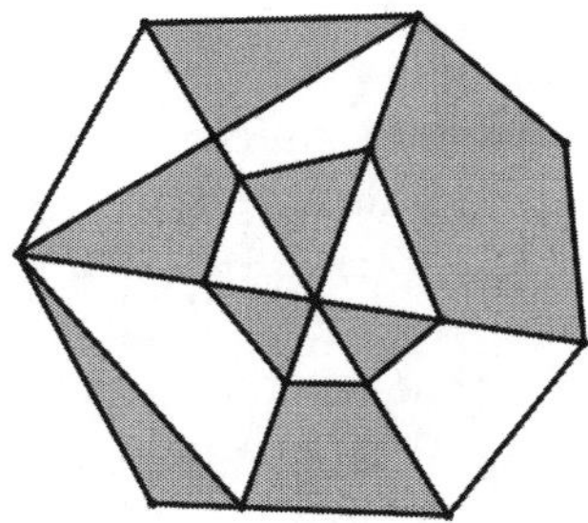

Figure 3.48. A map needing just two colors.

Theorems 3.10 and 3.11 give some conditions that require a map to use four colors or enable us to use only two colors. What about maps with three colors or more than four?

Problem 3.7 (a). Explore conditions that require three colors or more than four colors for maps.

Problem 2.7 (b). (repeated) Design as many different types of frieze patterns as you can, where two types are different when one has a symmetry the other doesn't have.

Recall that there are four possible symmetries besides horizontal translations (T), namely vertical mirrors (V), 180° rotations (R), horizontal mirrors (H), and

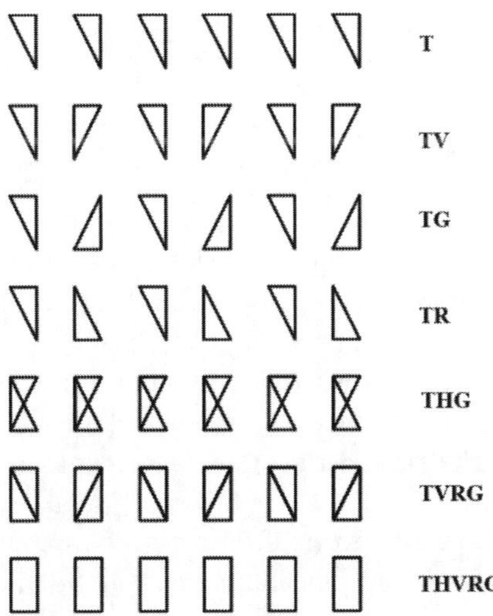

Figure 3.49. The seven types of frieze patterns.

glide reflections (G). Each of these four either is or isn't a symmetry of a frieze pattern. You might think with four options that there could be $2^4 = 16$ possible types of frieze patterns. However, if a pattern has a horizontal mirror and horizontal translations, it has to have glide reflections, which are built by combining these symmetries. Also, if a design has a 180° rotation and a glide reflection, it has to have a vertical mirror. There are other connections as well. Figure 3.49 gives the seven possible types of one-color frieze patterns. Three different mathematicians independently proved this classification of frieze patterns in the same year.

Theorem 3.12. (Niggli, Polya, Speiser, 1924) There are exactly seven different types of frieze patterns.

Proof. See [29, 276].   □

Exercise 3.3. Artists in many cultures have made frieze patterns from strands giving a layered effect, as in Figure 3.50. Classify the frieze pattern types of the patterns in Figure 3.50. Try your hand at designing other such patterns, including some with different symmetries.

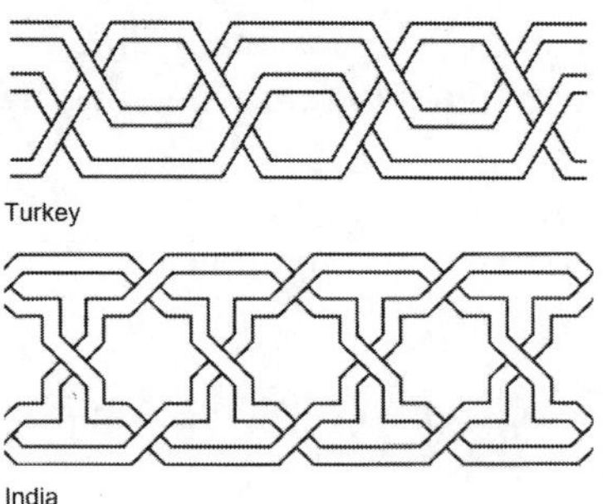

Figure 3.50. Layered frieze patterns from Turkey and India.

Problem 3.7 (b). The definition of a frieze pattern requires that the design repeat at set intervals, which makes the pattern discrete. Find examples continuous frieze patterns with different symmetry types.

Problem 2.7 (c). (repeated) Investigate two-color frieze patterns. Look for as many different types as you can, listing the types of all symmetries and the types of color-preserving symmetries. Hint: There are seventeen possible types of two-color frieze patterns.

Figure 3.51 shows how the seven different groups are related to one another—a group is a subgroup of a group placed above it provided there is a path always going up from the lower one to the upper one. The color-preserving subgroup of a two-color frieze pattern has to be as lower or lower in Figure 3.51 than the entire color group of symmetries. For instance, if the entire group has type TVRG,

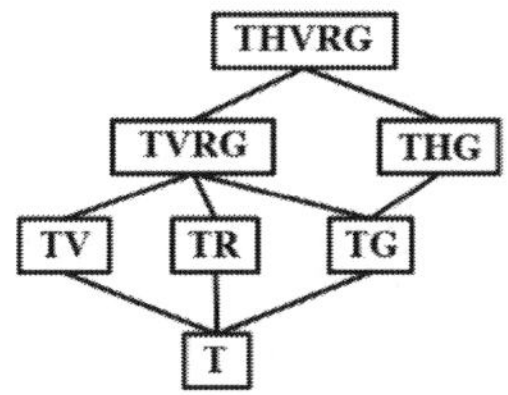

Figure 3.51. The subgroup relations of the seven types of frieze patterns.

there are five possibilities for the color-preserving subgroup, namely TVRG, TV, TR, TG, and T. There are in total 22 possibilities. However, not all of these can occur with two-color frieze patterns. Some require three or four colors. Figure 3.52 gives examples of the seventeen types of two-color frieze patterns.

Exercise 3.4. Classify the two-color frieze patterns in Figure 3.53.

Problem 3.7 (c). Find the five pairs of groups that can't be realized as two-color frieze patterns. Design frieze patterns using three or four colors for four of those types.

While frieze patterns have translations along a line, *wallpaper patterns* are patterns in the plane with translations along nonparallel lines. Figure 3.54 and 3.55 illustrate (parts of) two wallpaper patterns with different symmetry groups. The Spanish pattern in Figure 3.54, in addition to translations, has rotations of 90°, 180, and 270°. (Because the white designs are in layers, there are no reflections.) The design from Borneo has vertical and horizontal reflections and rotations of 180°. In 1891, the Russian mathematician and crystallographer Evgraf Fedorov (1853–1919) proved there were seventeen types of wallpaper patterns. (There are forty-six types of two-color wallpaper patterns, too many for us to reasonably explore. See [18, 408–413] for examples of these patterns.) Also in 1891, Fedorov proved the classification of the 230 three-dimensional crystal groups, needed in classifying chemical crystals.

Problem 3.7 (d). Look for examples of (some of) the seventeen different types of wallpaper patterns.

Problem 2.7 (d). (repeated) Is there a regular three-coloring and a regular four-coloring of the tessellation in Figure 2.31?

The left part of Figure 3.56 illustrates one way to give a regular coloring of the tessellation with three colors. It also suggests that we can find regular colorings with any number of colors for this tessellation with different colors going down different diagonals. The right part of Figure 3.56 gives a different arrangement for a regular four-coloring of the triangles.

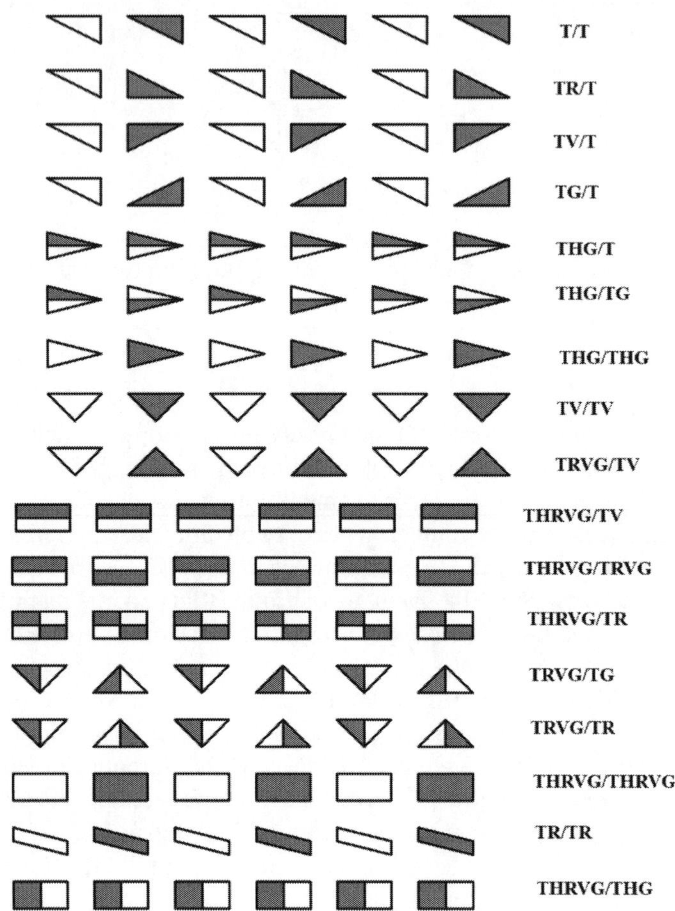

Figure 3.52. The seventeen types of two-color frieze patterns.

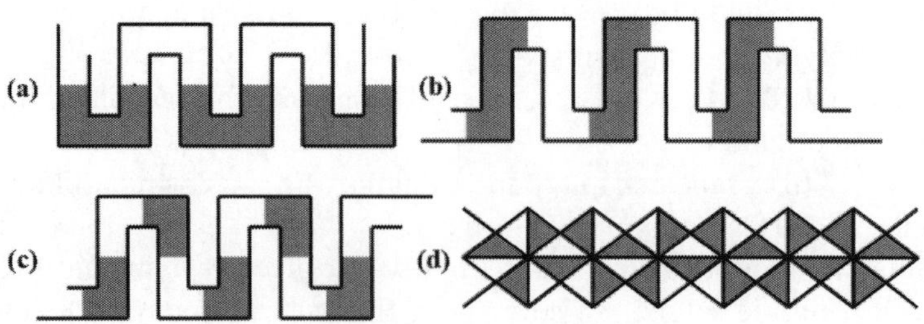

Figure 3.53. Two-color frieze patterns.

Figure 3.54. A wallpaper pattern from Spain.

Figure 3.55. A wallpaper pattern from Borneo.

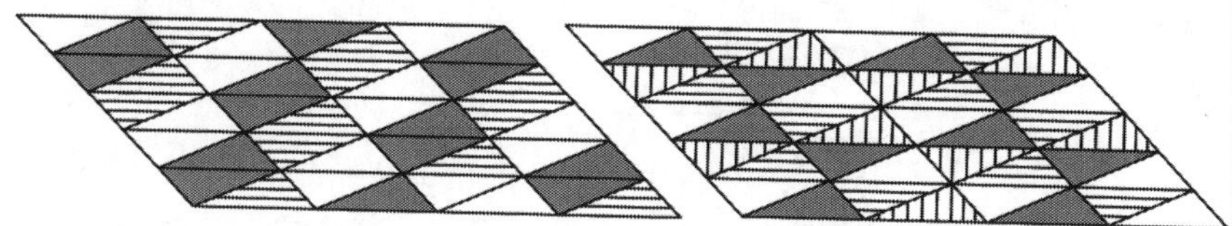

Figure 3.56. A triangular tessellation with regular three- and four-colorings.

Problem 2.7 (e). (repeated) Make tessellations and regular colorings using each of the quadrilaterals in Figure 2.32.

There are many ways to tessellate with rectangles, rhombi, and parallelograms, but just one way for a general quadrilateral. There can be more than one regular coloring for a tessellation. Figure 3.57 (a), (b), (c), and (d) gives one sample for each.

Problem 3.7 (e). Can every convex pentagon give a tessellation of the plane? If not, find some convex pentagons that will tessellate the plane. Repeat with hexagons.

We can shift back to the finite case, but consider regular colorings for three-dimensional polyhedra. Because the five regular polyhedra have interesting symmetry groups, they are natural choices for regular colorings.

Problem 3.7 (f). Find some regular colorings of the faces of the five regular polyhedra, shown in Figure 3.32, repeated here. The tetrahedron has four triangles, the cube has six squares, the octahedron has eight triangles, the dodecahedron has twelve pentagons, and the icosahedron has twenty triangles.

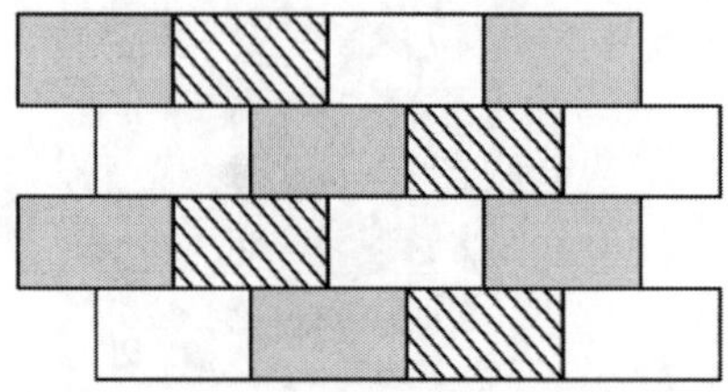

(a) A three-coloring of a tessellation with rectangles.

(b) A four-coloring of a tessellation with rhombi.

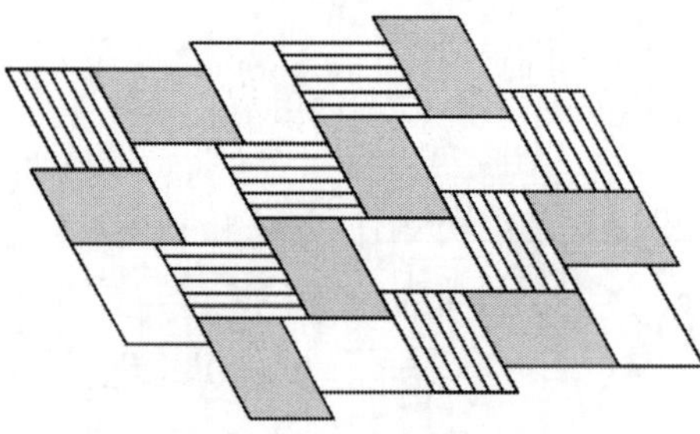

(c) A three-coloring of a tessellation with parallelograms.

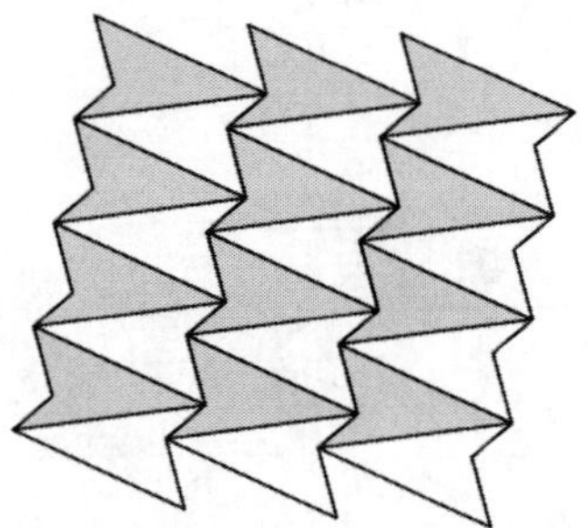

(d) A two-coloring of a tessellation with quadrilaterals.

Figure 3.57.

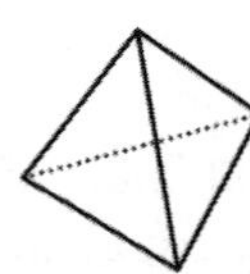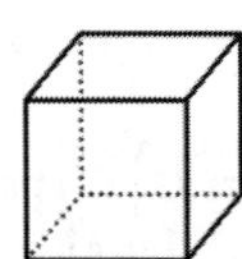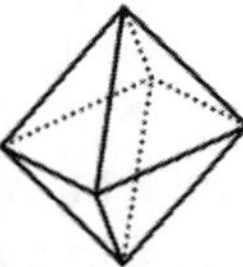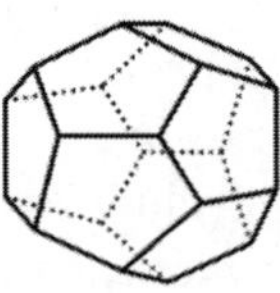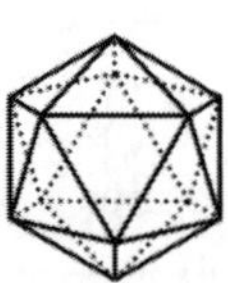

Figure 3.32. (repeated) The five regular polyhedra.

3.6 Voronoi Diagrams

Problem 2.8 (a). (repeated) Is every region of a Voronoi diagram convex?

Yes. The perpendicular bisector between two sites creates two half planes. (Figure 3.58 shades the half plane containing A.) We build Voronoi regions by intersecting the half planes including a given site, as Figure 3.59 indicates. These half planes are convex and, as Theorem 3.13 demonstrates, the intersection of

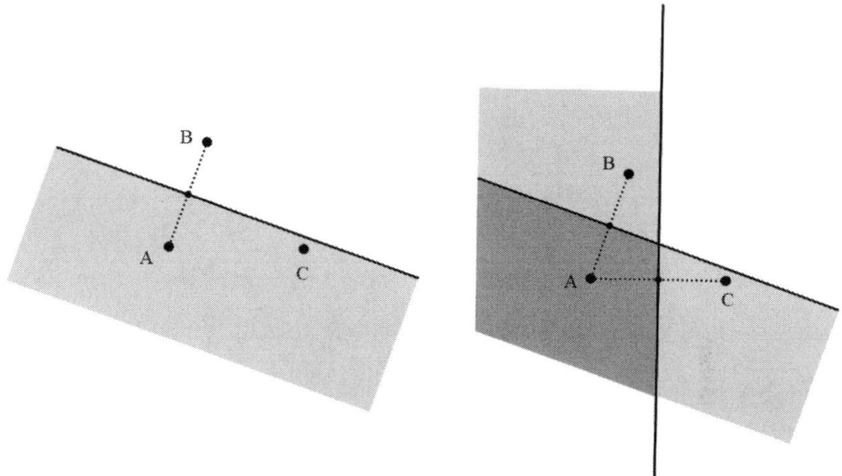

Figure 3.58. The shaded half plane containing A for the perpendicular bisector of $\overline{AB}$.

Figure 3.59. The intersection of two half planes containing A.

convex sets is also convex. (The proof holds even if there are infinitely many convex sets to intersect.)

Theorem 3.13. The intersection of convex sets is convex.

Proof. Let S_1, S_2, … be convex sets and A and B be any two points in the intersection of all of these sets. For each S_i, the segment $\overline{AB}$ is a subset of S_i by definition of convex. Since this segment is in each of the sets S_i, it is in their intersection. That is, the intersection is convex. $\square$

Problem 3.8 (a). Explore generalizing Voronoi diagrams to three-dimensional Euclidean space or even higher dimensions. Will the regions still be convex?

Problem 2.8 (b). (repeated) What is the shape of the Voronoi regions if the sites form a repeating lattice with translations in different directions, as in Figure 2.37?

As Figure 3.60 suggests, the Voronoi diagram for a repeating lattice of sites can be a tessellation of hexagons. In special cases, such as the sites forming a rectangular lattice, the regions are parallelograms, as in Figure 3.61. Fedorov found these possibilities in 1885.

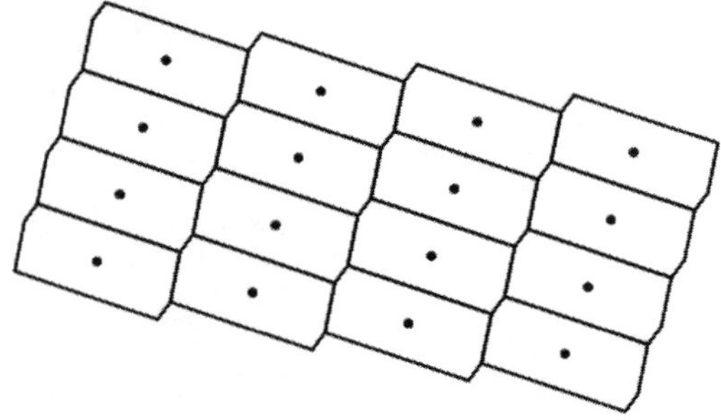

Figure 3.60. A Voronoi diagram for a repeating lattice of sites.

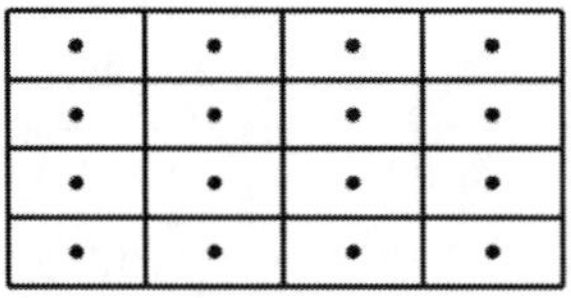

Figure 3.61. Voronoi diagram for a rectangular lattice of sites.

Problem 2.8 (c). (repeated) Devise a method to find the Voronoi diagram for a set of *n* points in the Euclidean plane.

An inefficient method would be to construct all perpendicular bisectors and their intersections. Then omit the unneeded intersections, segments and rays. A more systematic way starts from one side, say the bottom and moves up. As each new point is included, we adjust the developing Voronoi diagram. A problem with this is that some provisional line segments and rays may need to be shortened or even eliminated. Suppose in Figure 3.62 that we have moved up from the bottom enough to consider points A, B, and C, but not D and E. The Voronoi diagram at this point would consist of the three solid rays shown. However, adding in D and E gives the diagram in Figure 3.63 in which the upward pointing ray of Figure 3.62 is entirely replaced.

Computer algorithms that automate finding Voronoi diagrams don't find all the provisional segments and rays at every step of the way. They compute the parts that won't change as more points are added. In our example, if the computer has computed up to the line k in Figure 3.64, it could be sure of the lower part of the diagram, as shown. As the line k moves up, more of the lower rays would appear. (See [11, 104–107] for more on algorithms for constructing Voronoi diagrams.)

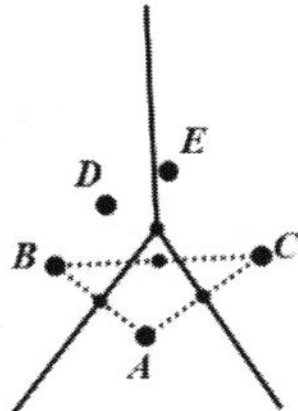

Figure 3.62. Voronoi diagram based on A, B, and C.

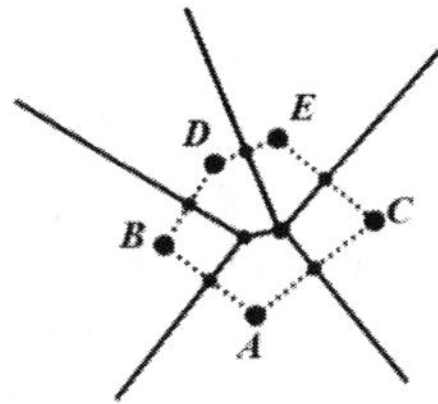

Figure 3.63. Complete Voronoi diagram.

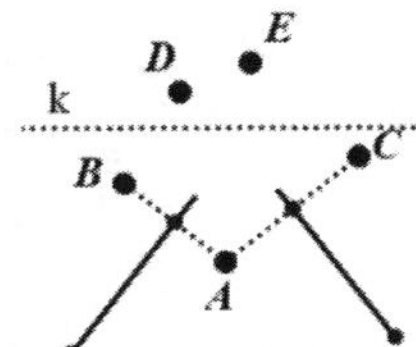

Figure 3.64. Partial Voronoi diagram based on A, B, and C.

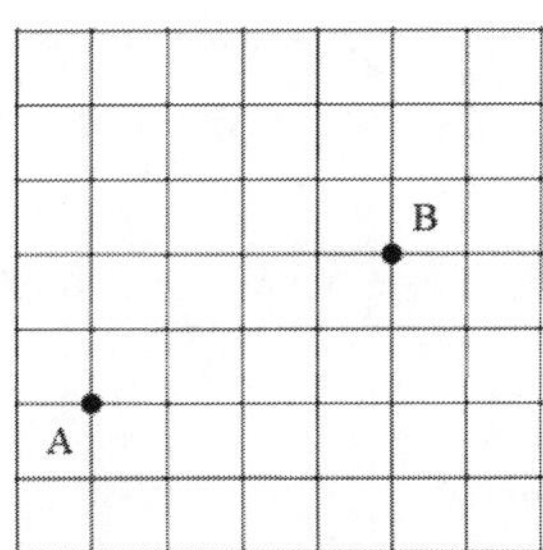

Figure 3.65. Fire stations in a hypothetical town.

In many practical applications, the usual distance doesn't fit. Suppose in Figure 3.65 an idealized town with streets as indicated has two fire stations at points A and B. The fire trucks have to go on the streets, so the taxicab metric gives a better way to determine which parts of the city are closer to station A or closer to B. As Exercise 3.5 will show, the two regions are not always separated by a straight line, let alone the perpendicular bisector.

Definition. The *Voronoi boundary* in the taxicab metric between two points is the set of points the same distance from both points.

Exercise 3.3. Find the Voronoi boundary between points A and B in Figure 3.65 using the taxicab metric.

Problem 3.8 (b). In Figures 3.66 and 3.67, determine the Voronoi boundary between C and D and the Voronoi boundary between E and F. Describe the possible shapes of Voronoi boundaries in the taxicab metric.

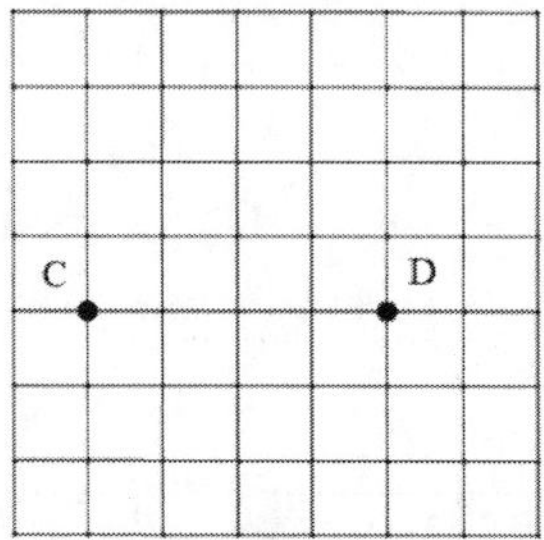

Figure 3.66. A special case for a Voronoi boundary.

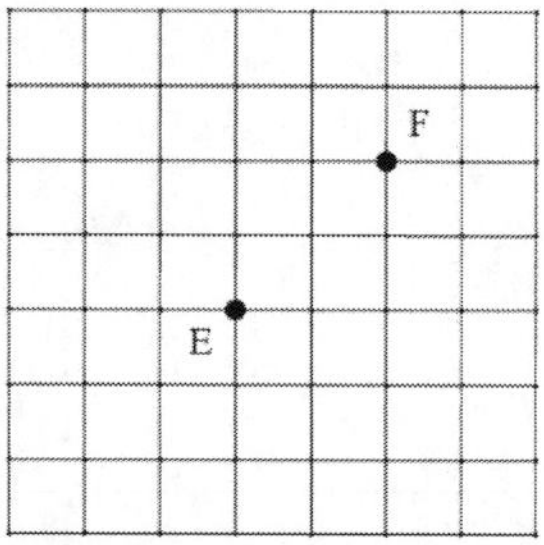

Figure 3.67. A different special case for a Voronoi boundary.

Problem 3.8 (c). Find the Voronoi boundaries in the taxicab metric if the sites are A = $(1, 1)$, B = $(3, 7)$, C = $(7, 5)$, D = $(5, 9)$, and E = $(10, 4)$. Compare with the Voronoi diagram for these points using the usual Euclidean distance.

4

Final Explorations and Connections

4.1 Lines and Regions

Problem 3.1 (a). (repeated) Try to find arrangements of 4, 5, and other values of n lines to get all the values for the number of regions they determine between $2n$ and $\frac{n^2+n+2}{2}$.

With four lines, it is possible to get all the values between $2n = 8$ and $\frac{n^2+n+2}{2} = 11$, as Figure 4.1 indicates. For a larger number of lines, we can't get all values between $2n$ and $\frac{n^2+n+2}{2}$ regions. For instance, Figure 4.2 gives arrangements of five lines with 12, 13, 14, and 15 regions. However, no arrangement of five lines determines exactly 11 regions.

The analysis based on intersections from Section 3.1 becomes more complicated with five or more lines. As noted there, no intersections happen with

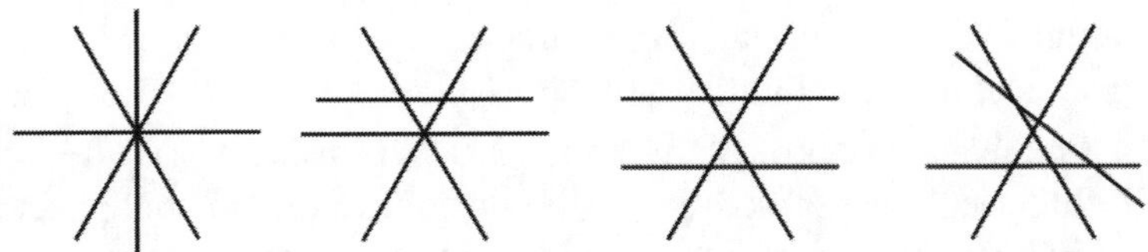

Figure 4.1. Arrangements of four lines giving 8, 9, 10 , and 11 regions.

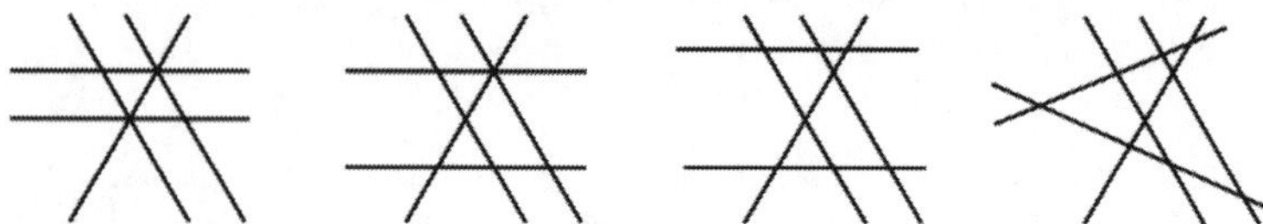

Figure 4.2. Arrangements of five lines giving 12, 13, 14, and 15 regions.

parallel lines and give six regions. All five lines through a common point of intersection determines ten regions. We can't have just two or three points of intersections. There are two ways to have four points of intersection: Four parallel lines and one line intersecting them give ten regions. Alternatively, as in the drawing on the left of Figure 4.2, we can have two sets of parallel lines with the fifth line through two of their intersections. But then we jump to twelve regions. Once we have more intersections, we get more regions, until with all $\binom{5}{2} = 10$ intersections we have the general position and 16 regions.

We can generalize the first three arrangements in Figure 4.1. With n lines all through the same point there will be $2n$ regions. With $n - 1$ lines through a point and one parallel to one of these, there are $3n - 3$ regions. With $n - 2$ lines through a point and two parallel lines not parallel to any of the previous ones, there are $4n - 6$ regions. Here are some other possibilities. With $n - 1$ lines through a point and a line not parallel to any of them nor through that point, we have $3n - 2$ regions. With $n - 2$ lines through a point and two more parallel to one of them, we have $4n - 8$ regions. Also, k parallel lines intersecting $n - k$ parallel lines determine $(k + 1)(n - k + 1)$ regions.

Problem 3.1 (b). (repeated) What are the maximum and minimum number of regions in four-dimensional space determined by n three-dimensional hyperplanes? Generalize for five-dimensional space and higher dimensions.

In any number of dimensions, the minimum occurs with n parallel hyperplanes, giving $n + 1$ regions. For the maximum, let's extend the reasoning we did for the three-dimensional case and employ some notation to make that reasoning clearer.

Definition. Let $M_k(n)$ be the maximum number of regions determined by n $(k - 1)$-dimensional hyperplanes in k dimensions.

For instance, we have found $M_2(n) = M(n) = \frac{1}{2}n^2 + \frac{1}{2}n + 1$ and $M_3(n) = \frac{1}{6}n^3 + \frac{5}{6}n + 1$. We are seeking $M_4(n)$, $M_5(n)$, and possibly $M_k(n)$. How do hyperplanes intersect? Let's look at lower dimensions to reason analogously. Two (nonparallel) lines intersect in a point and two nonparallel planes intersect in a line. We can say this more generally: two k-dimensional objects (hyperplanes) intersect in a $(k - 1)$-dimensional object (hyperplane). What about three in general position? For three lines, there is no common intersection. With three planes in general position, there is a point—we have dropped two dimensions. So, we expect three k-dimensional hyperplanes in general position to intersect in a $(k-2)$-dimensional hyperplane. The more hyperplanes in general position we intersect, the lower the dimension of the intersection is. Four k-dimensional hyperplanes in general position would then intersect in a $(k-3)$-dimensional hyperplane (as long as $k-3 \geq 0$). We keep dropping a dimension with additional hyperplanes until we get to a point, which happens with $k+1$ of these k-dimensional hyperplanes.

Finally in a high enough dimension, $k+2$ or more k-dimensional hyperplanes in general position don't have a common intersection. For instance, when $k=1$, we found that two nonparallel lines in the plane meet in a point and three general lines have no common intersection. For $k=2$, three planes in general position in three-dimensional space meet in a point and four have no common intersection. We want to count the number of regions determined by these intersecting hyperplanes in $(k+1)$-dimensional space. As we add an additional k-dimensional hyperplanes in general position up to $k+1$ of them, we double the number of regions. Table 4.1 gives the start of the sequence of maximum regions.

Table 4.1. The maximum number of regions for n k-dimensional hyperplanes in $k+1$ dimensions.

hyperplanes	1	2	3	...	$k+1$
maximum regions	2	4	8	...	2^{k+1}

To see how we extend the sequence beyond $k+1$ hyperplanes in $k+1$ dimensions, let's revisit the three-dimensional reasoning. With three planes in general position we have eight regions. These three planes divide a fourth plane into $M(3) = M_2(3) = 7$ regions. Each of these split a three-dimensional region in half, adding that many new regions. That is, $M_3(4) = M_3(3) + M_2(3) = 8 + 7 = 15$. When we added a fifth plane, the four current planes divided it into $M_2(4) = 11$ regions. These in turn split that many spacial regions, giving $M_3(5) = M_3(4) + M_2(4) = 15 + 11 = 26$. In general, $M_3(n+1) = M_3(n) + M_2(n)$.

Let's generalize this to $k+1$ dimensions. We add a $(k+2)^{\text{nd}}$ k-dimensional hyperplane, called H, in general position in $k+1$ dimensions. Then H intersects the other hyperplanes in $(k-1)$-dimensional hyperplanes. These $(k-1)$-dimensional hyperplanes split H into $M_k(k+1)$ regions (which have dimension k) since the hyperplanes are in general position. In turn, each of these k-dimensional regions splits a $(k+1)$-dimensional region into two. That is, $M_{k+1}(k+2) = M_{k+1}(k+1) + M_k(k+1)$. In general for $n > k+1$, $M_{k+1}(n+1) = M_{k+1}(n) + M_k(n)$.

Table 4.2. Initial values of $M_2(n)$, $M_3(n)$, $M_4(n)$, and $M_5(n)$.

n	1	2	3	4	5	6	7
$M_2(n)$	2	4	7	11	16	22	29
$M_3(n)$	2	4	8	15	26	42	64
$M_4(n)$	2	4	8	16	31	57	99
$M_5(n)$	2	4	8	16	32	63	120

Table 4.2 gives the start of the sequences for $M_2(n)$, $M_3(n)$, $M_4(n)$, and $M_5(n)$.

We can solve equations, as in the solution to Problem 2.1 (b), or use the difference equations described in Section 3.1 to find formulas for $M_4(n)$ and $M_5(n)$. They are $M_4(n) = (n^4 - 2n^3 + 14n^2 + 7n + 24)/24$ and $M_5(n) = (n^5 - 5n^4 + 25n^3 + 5n^2 + 94n + 120)/120$. Mathematicians have developed another option. The *On-line Encyclopedia of Integer Sequences* (abbreviated OEIS) provides a valuable tool for anyone investigating sequences. You can access this powerful and free web source at `https://oeis.org`. To use it once you are at the site, put the first few entries of a sequence you are investigating into OEIS, and it will list sequences that others have studied having those numbers in that order. For instance, if you put $2, 4, 7, 11$ into OEIS, the first sequence it suggests is $A000124$, which represents our sequence of the maximum number of regions in a plane determined by n lines. It gives lots of information about that sequence, including the formula we have found. (For that same input, OEIS also gives other sequences. For instance, $A004250$, which starts $0, 0, 1, 2, 4, 7, 11, 17, 25$, which counts the number of different ways to partition the integer n into at least three positive integers.) If you put in a sequence OEIS doesn't already have, it will ask you to send in the information you have about the sequence. The people running OEIS are always looking for new sequences and new mathematical connections.

The sequence $A000125$ matches $M_3(n)$, the maximum number of three-dimensional regions determined by n planes. Here are some of the others we have considered:

$A014206$, $C(n)$, the maximum number of regions in the plane determined by n circles.

$A000127$, $M_4(n)$, the maximum number of four-dimensional regions determined by n three-dimensional hyperplanes.

$A006261$, $M_5(n)$, the maximum number of five-dimensional regions determined by n four-dimensional hyperplanes.

$A008859$, $M_6(n)$, the maximum number of six-dimensional regions determined by n five-dimensional hyperplanes.

Problem 3.1 (c). (repeated) Design a Venn diagram for four convex sets that has all sixteen possible regions.

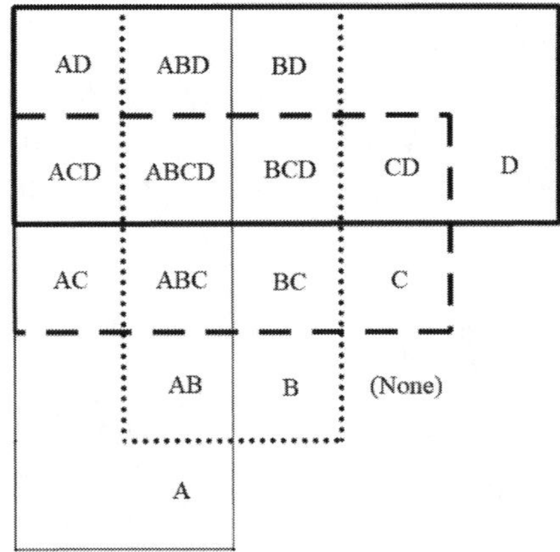

Figure 4.3. Venn diagram for four sets.

We illustrate one of many possibilities. Figure 4.3 provides one Venn diagram with four intersecting rectangles and four categories. The labels of the regions indicate all the categories present in that region.

Problem 3.1 (d). (repeated) What is the maximum number of regions in space determined by n spheres?

Recall the reasoning to solve Problem 2.1 (b) by going from regions determined by lines to regions determined by planes. This approach also works to go from regions determined by circles to regions determined by spheres. Table 4.3 suggests that the maximum number of regions determined by $n + 1$ spheres is the sum of the maximum numbers for n spheres and n circles. We can write this recursive pattern as $S(n + 1) = S(n) + C(n)$, where $S(n)$ is the maximum number of regions for n spheres and $C(n)$ is the maximum number of regions for n circles. Since the formula for the circles was second degree, you shouldn't be surprised that the formula for n spheres is third degree, $S(n) = \frac{1}{3}n^3 - n^2 + \frac{8}{3}n$. The sequence is in the OEIS, sequence number $A046127$. Our reasoning here will extend to higher dimensions as well. For example, let $H(n)$ be the maximum number of regions for n four-dimensional hyperspheres. Then $H(n + 1) = H(n) + S(n)$,

Table 4.3. Maximum number of regions determined by n circles or spheres.

n	1	2	3	4	5	6
circles	2	4	8	14	22	32
spheres	2	4	8	16	30	52

$H(n) = \frac{1}{12}n^4 - \frac{1}{2}n^3 + \frac{23}{12}n^2 - \frac{3}{2}n + 2$, and in OEIS the sequence is $A059173$. You might suspect (correctly!) that many mathematicians enjoy playing in higher dimensions—there are so many intriguing questions to explore.

Problem 3.1 (e). (repeated) What is the maximum number of regions in the plane determined by two nonconvex polygons each with the same even numbers of sides?

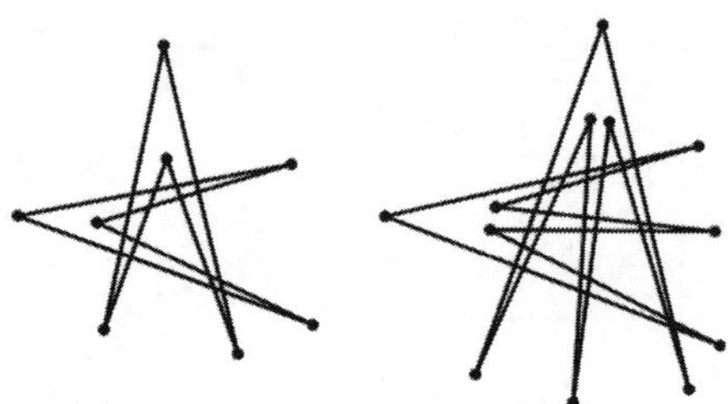

Figure 4.4. Regions determined by two nonconvex quadrilaterals and two nonconvex hexagons.

The maximum number of intersections of two nonconvex $2n$-gons form the sequence $A005899$ in OEIS, with the formula $4n^2 + 2$ and illustrated in Figure 4.4. Table 4.4 gives a few of these values.

Table 4.4. Maximum number of regions determined by two $2n$-gons.

n	2	3	4	5
regions	18	38	66	102

Reasoning: Each edge of one polygon can intersect an edge of the other polygon at most once. So the greatest number of intersections and so regions occurs when each edge of each polygon intersects each edge of the other one. In Figure 4.5, the shaded central quadrilateral has $(2n-1)^2$ regions, there are $2(n-1)$ interior triangles embedded in bigger triangles, there are $2(n+1)$ exterior triangles, and finally there is the outside. That gives $(2n-1)^2 + 2(n-1) + 2(n+1) + 1 = 4n^2 + 2$.

The OEIS also tells us this sequence gives the number of points on the surface of an octahedron when there are n points on an edge not counting the vertices. (See Figures 4.6 and 4.7.) Unexpected connections such as this provide another of the many pleasures of mathematics.

For the octahedron, we always have the six vertices, which corresponds to $n = 0$, instead of $n = 1$ for the intersecting polygons. This will shift the expression for the formula by one. The twelve edges contribute $12n$ points (besides the

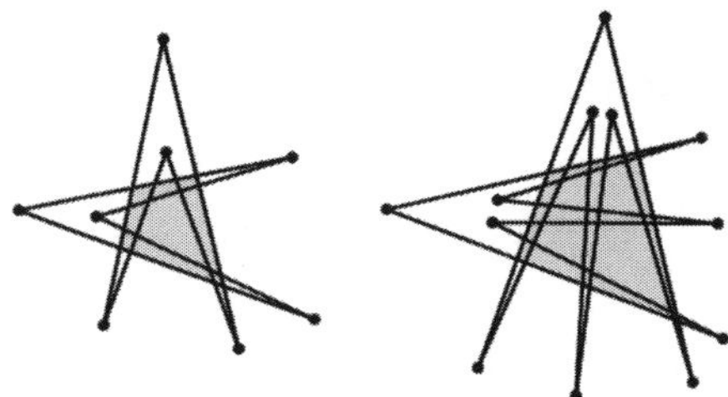

Figure 4.5. Regions of intersecting nonconvex polygons.

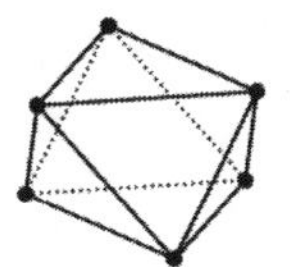 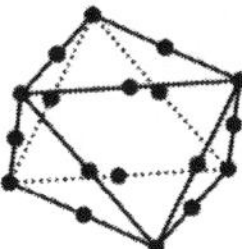 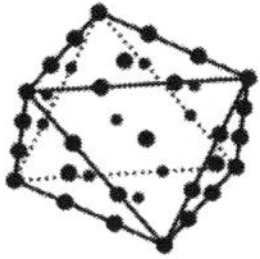   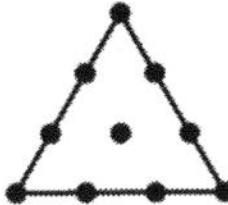 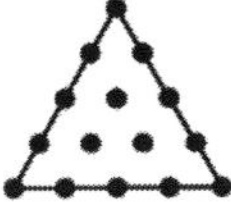

Figure 4.6. Number of points on the surface of an octahedron with n points on an edge.

Figure 4.7. Individual face of octahedron.

vertices) and the eight faces each add a triangular number, $8\frac{(n-1)n}{2}$ (besides those on edges or vertices). The total is thus $6 + 12n + 4n^2 - 4n = 4(n+1)^2 + 2$.

4.2 Diagonals and Triangulations

Problem 3.2. (repeated) How can we count the number of interior diagonals of a convex polyhedron?

Each face of a convex polyhedron is a convex polygon, so we can count the surface diagonals provided we know the number of triangular faces, quadrilateral faces. etc. The answer to Problem 2.4 (c) led us to Euler's formula for convex polyhedra: $V - E + F = 2$, where V is the number of vertices, E the number of edges, and F the number of faces. Let S be the number of surface diagonals and I the number of interior diagonals. Then $\binom{V}{2} = \frac{V(V-1)}{2} = E + S + I$ since in a convex polygon every line segment between two vertices must be one of these three options.

Let F_3 be the number of triangular faces, F_4 the number of quadrilateral faces, and so on. Then $S = 0F_3 + 2F_4 + 5F_5 + 9F_6 + \ldots$. We can write this sum more succinctly with " sigma" notation. The Greek letter $\sum$ (Sigma) stands for sum and we use a subscript to describe the first term in the sum and a superscript for the last term and each term is given in terms of a variable. For instance, in the term $9F_6$, the nine diagonals of a hexagon come from the formula $\frac{6(6-3)}{2}$. The general term for faces with i vertices is $\frac{i(i-3)}{2}F_i$. Then the total number of surface diagonals is $S = \sum_{i=3}^{k} \frac{i(i-3)}{2}F_i$, where k is the number of vertices of the face with the most vertices. We can go one step further to give an expression for E in terms of the F_i values: $2E = \sum_{i=3}^{k} iF_i$. Theorem 4.1 summarizes this reasoning.

Example. A cuboctahedron, illustrated in Figure 4.8, has eight triangles and six squares for faces. We have $V = 12$, $E = 24$, $F = 14$, $F_3 = 8$, and $F_4 = 6$. We find that $\binom{V}{2} = 66$. Also $S = 2 \cdot 6 = 12$. Then $66 = 24 + 12 + I$, giving $I = 30$. Because of the symmetry of this polyhedron, we can find I more easily directly: Each vertex has four edges and two surface diagonals, forcing five interior diagonals from it. There are twelve vertices, seemingly giving $5 \cdot 12 = 60$ interior diagonals. But we have counted each such diagonal twice, halving the number of interior diagonals to 30.

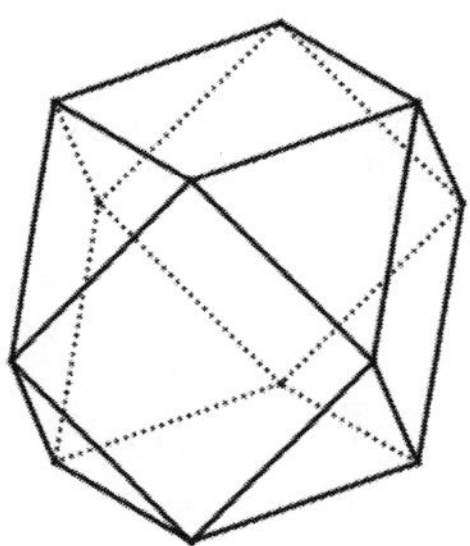

Figure 4.8. A cuboctahedron.

Theorem 4.1. If F_i is the number of faces with i vertices for $3 \le i \le k$ and V is the number of vertices of a convex polygon, then the number of internal diagonals is $I = \frac{1}{2}(V(V-1) - \sum_{i=3}^{k} i(i-2)F_i)$.

Proof. We combine the reasoning prior to the example: $\frac{V(V-1)}{2} = E + S + I = \frac{1}{2}\sum_{i=3}^{k} iF_i + \frac{1}{2}\sum_{i=3}^{k} i(i-3)F_i + I$ and solve for I. $\qquad\square$

Problem 3.3. (repeated) For each k construct a (possibly nonconvex) polyhedron with exactly k interior diagonals.

We don't need the polyhedron to be nonconvex if we want just one interior diagonal: a triangular bipyramid has five vertices, six faces that are all triangles, nine edges, no surface diagonals, and thus one interior diagonal. But Figure 4.9 pushes one of the pyramids inside the other to make the bipyramid nonconvex, which in turn suggests a way to get polyhedra with a specified number of interior

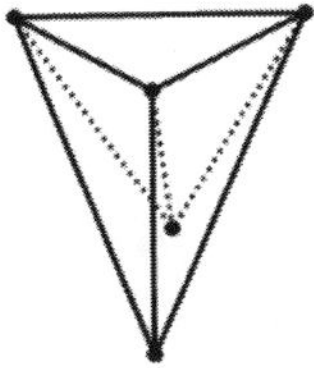 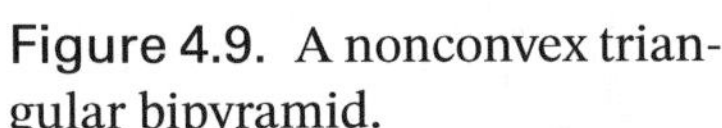

Figure 4.9. A nonconvex triangular bipyramid.

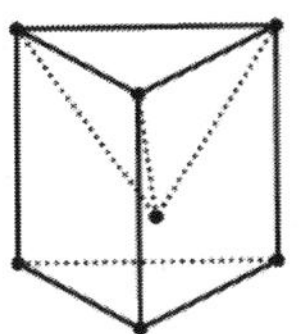 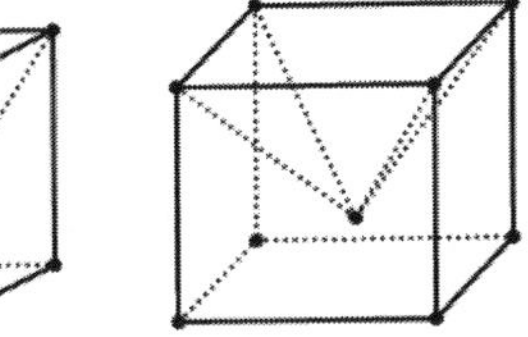

Figure 4.10. Prisms minus pyramids.

diagonals. Figure 4.10 illustrates the idea for k interior diagonals, where $k \geq 3$. We take an k-gonal prism and dig out an k-gonal pyramid. The apex of the missing pyramid must be low enough to eliminate interior diagonals from a bottom vertex to a top vertex.

There are $2k + 1$ vertices, $4k$ edges, k rectangular faces, k triangular faces and one k-gon on the bottom. This gives $2k + \frac{k(k-3)}{2}$ surface diagonals. There are k interior diagonals from the apex of the missing pyramid to the bottom polygon. The other potential surface and interior diagonals would have some of their length going through the missing pyramid. Let's count these. There are $\frac{k(k-3)}{2}$ missing surface diagonals that would have been be on the top polygon. There are $k(k - 3)$ interior diagonals that would have gone from a bottom vertex to a top vertex. The sum of all the edges, existing surface and interior diagonals, and missing diagonals is $4k + 2k + \frac{k(k-3)}{2} + k + \frac{k(k-3)}{2} + k(k - 3) = 2k^2 + k$. Finally, with $2k + 1$ vertices, we get the same number, $\binom{2k+1}{2} = \frac{(2k+1)2k}{2} = 2k^2 + k$, of possible line segments connecting all pairs of them. So there are indeed exactly k interior diagonals. We leave the special case of exactly two interior diagonals to the reader.

Problem 3.4 (a). (repeated) Draw a variety of polygons with lattice points as vertices and find their areas, the number of interior lattice points, and the number of lattice points on their edges. Is there a relation among these three numbers?

Let's start with some rectangles, whose areas and interior and boundary points are easy to count, as in Figure 4.11. Table 4.5 gives their areas and the number of interior and boundary points.

Table 4.5. Area, interior points, and boundary points of rectangles.

rectangle size	1×1	2×1	3×1	3×2	4×2	3×3	4×3
area	1	2	3	6	8	9	12
interior	0	0	0	2	3	4	6
boundary	4	6	8	10	12	12	14

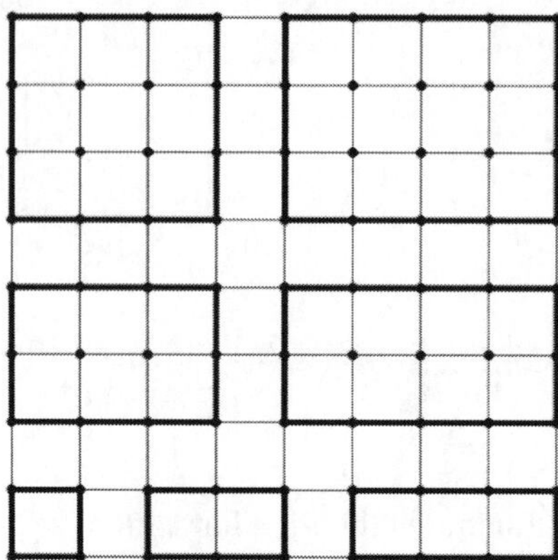

Figure 4.11. Selected rectangles with lattice points.

From the three rectangles with height 1, it appears that the area A goes up 1 when the boundary B points increase by 2. The actual areas of these rectangles is 1 less than half the number of boundary points, suggesting $\frac{1}{2}B - 1$ for the area. Comparing the 4×2 and 3×3 rectangles, it appears that the area goes up 1 when the interior points increase by 1. Together these suggest the formula $A = \frac{1}{2}B + I - 1$, which fits with the other rectangles. Of course, there is no guarantee that this formula will hold for more irregular shapes than rectangles. The polygon in Figure 3.16 had area 36.5 square units, 28 interior lattice points and 19 boundary lattice points. These values also fit our formula: $\frac{1}{2}(19) + 28 - 1 = 36.5$. Indeed, this is the formula Georg Pick found and proved.

We found that there were $n + 2k - 2$ triangles in a triangulation of a polygon with n vertices and k interior points. If we set $B = n$ and $I = k$ and divide our formula for the number of triangles by 2, we get Pick's theorem—a seemingly wonderful connection. In Figure 4.11, we can triangulate each of the rectangles using right triangles that are one unit on a side, which have area of $\frac{1}{2}$. That explains in this easy case why we divide our formula for the number of triangles by two to get the area of the polygon.

The hard part of the proof of Pick's theorem (Theorem 4.2) shows that the area of $\frac{1}{2}$ holds for every triangle whose vertices are lattice points and that have no other lattice points either on the boundary or in the interior. An example may illustrate why this is more challenging to prove in general. The triangle in Figure 4.12 has only its vertices as lattice points. It is awkward to find its area directly,

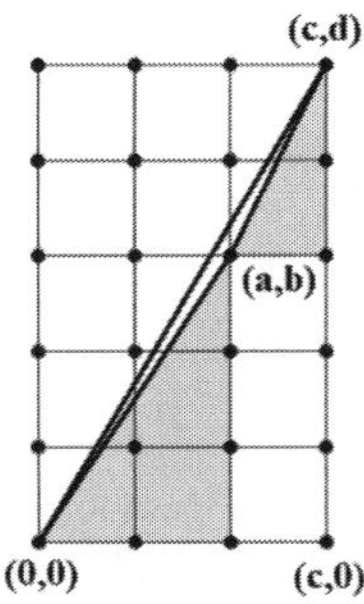

Figure 4.12. A triangle with area $\frac{1}{2}$.

but it is the difference of the areas of a big right triangle minus the areas of the shaded right triangles and the remaining rectangle. The area in general is

$$\frac{1}{2}cd - \left(\frac{1}{2}ab + \frac{1}{2}(c-a)(d-b) + (c-a)b\right) = \frac{1}{2}(ad - bc).$$

In Figure 4.12, $a = 2$, $b = 3$, $c = 3$, and $d = 5$. So the area is $\frac{1}{2}(2 \cdot 5 - 3 \cdot 3) = \frac{1}{2}$, as desired. You can try other long thin triangles and verify they have area $\frac{1}{2}$ provided their lattice points are just their vertices.

Theorem 4.2. (Pick's theorem, 1899). A polygon whose vertices are lattice points has area $\frac{B}{2} + I - 1$, where B is the number of lattice points on the polygon's boundary and I is the number of interior lattice points.

Proof. See [8, 209]. $\qquad\square$

Georg Pick was a distinguished Austrian mathematician living in what is now the Czech Republic, although he is remembered almost only for the theorem we have investigated. He had long retired when World War II broke out. Because he was Jewish, the Germans sent him to a concentration camp, where he died after two weeks.

Exercise 4.1. Find and prove a formula for the volume of a *lattice box* whose eight vertices are lattice points in terms of the number of lattice points on its edges, surfaces, and interior as well as its vertices. (See Figure 4.13.)

Unfortunately, there can't be a three-dimensional Pick's theorem for all lattice polyhedra. (Reeve [25] proves a more general theorem about lattice polyhedra than just boxes.) The formula for lattice boxes in Exercise 4.1 (in which all the terms are first degree, like Pick's theorem) doesn't work more generally. We'll show that there can be no such formula using some other polyhedra. The smallest lattice polyhedron is a triangular pyramid with area $\frac{1}{6}$, 4 lattice points for vertices, and no other lattice points. The smallest square pyramid has area $\frac{1}{3}$, 5 lattice points for vertices, and no other lattice points. A unit cube has eight lattice points for vertices and no other lattice points. (See Figure 4.14 for all three of these polyhedra.) Whatever coefficients there might be in general for the number of edge, surface, and interior lattice points, no coefficient for the number of vertices will work. Suppose, for a contradiction, that the formula was $aV + b$ for some constants a and b (plus other constants for the number of edges, surface, and interior lattice points). We'd have $4a + b = \frac{1}{6}$, $5a + b = \frac{1}{3}$, and $8a + b = 1$. The first two equations give $a = \frac{1}{6}$ and $b = \frac{-3}{6}$, but these are incompatible with the cube.

Exercise 4.2. Find the area of regular tetrahedron whose vertices are four vertices of a unit cube, no two of which are adjacent to each other. Explain how this and the smallest lattice triangular pyramid give another way to show the impossibility of a three-dimensional Pick's theorem.

Problem 3.4 (b). (repeated) Explore geometric ways that a computer might use to decide which triangulation might be the most realistic for the fourteen points in Figure 3.17 (reproduced below).

Boris Delaunay (1890–1980), a Russian mathematician, determined in 1934 how to find a triangulation that avoids narrow triangles. Computer scientists sometimes use this triangulation in computer graphics to make representations of geographical surfaces. As in Figure 3.17, they start with a set of coordinates of points. They use a triangulation to create a surface from the isolated points. Although there are exceptions, steep narrow valleys or ridges are unusual in nature. So a *Delaunay triangulation* generally gives a reasonable representation. Rather than give a computer algorithm, we investigate a geometrical way to make his triangulation. A qualifying triangle $\triangle P_i P_j P_k$ satisfies the property that no other point P of the set is inside the circle going through the vertices of the triangle. In

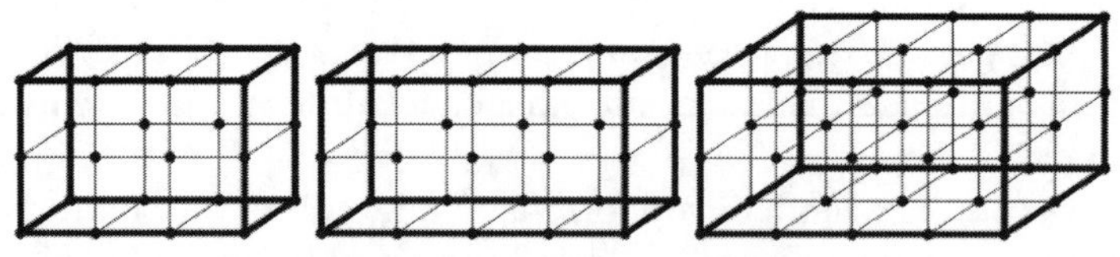

Figure 4.13. Lattice boxes.

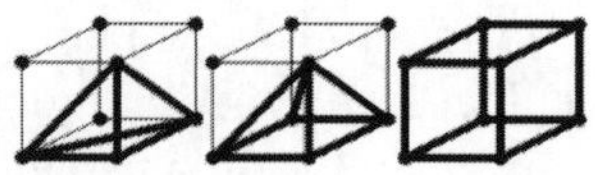

Figure 4.14. A smallest lattice triangular pyramid, a lattice square pyramid, and a lattice cube.

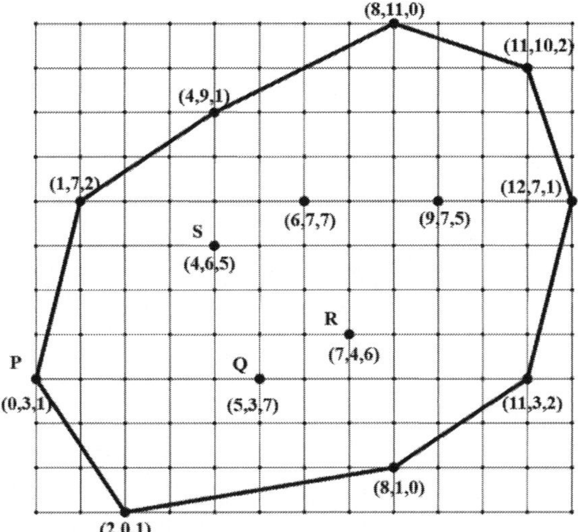

Figure 3.17. (repeated) A set of points to triangulate to best represent a three-dimensional surface.

the left part of Figure 4.15, the circles for triangles $\triangle$PQR and $\triangle$PRS contain the other point, so this choice can't be part of a Delaunay triangulation. On the right of Figure 4.15 the fourth point is outside of each of the circles for the triangles $\triangle$PQS and $\triangle$QRS. This distinction can be extended to give a triangulation. Figure 4.16 gives the Delaunay triangulation of the fourteen points of Figure 3.17 and Figure 4.17 shows the triangulated surface.

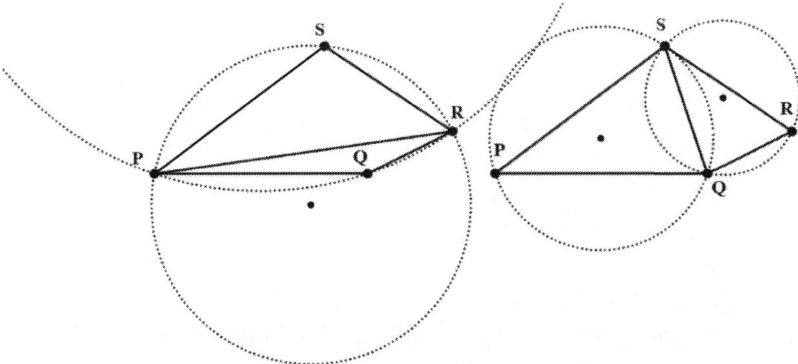

Figure 4.15. Determining whether the circles through the vertices of a triangle contain another point of the set.

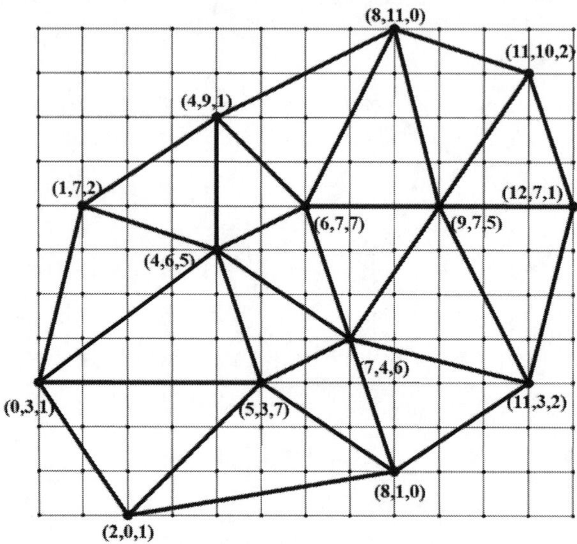

Figure 4.16. The Delaunay triangulation of Figure 3.17.

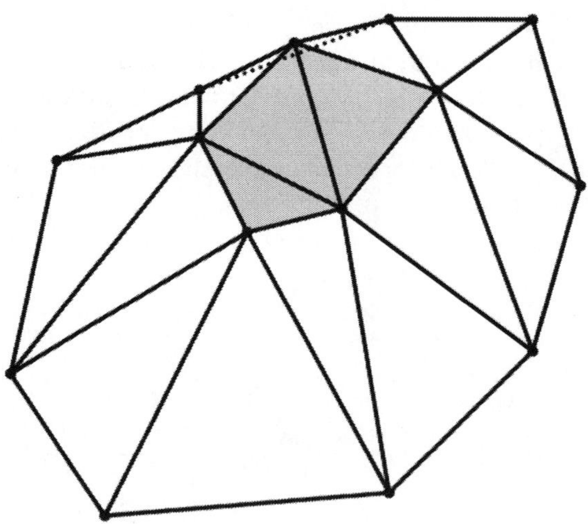

Figure 4.17. The surface determined by the Delaunay triangulation with the highest part of the surface shaded.

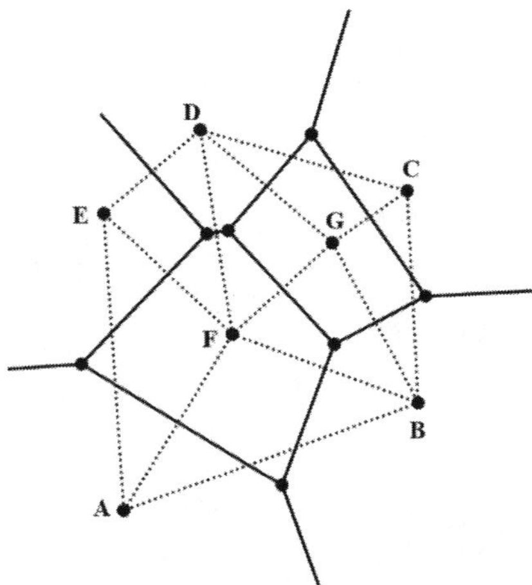

Figure 4.18. The Voronoi diagram and Delaunay triangulation
for the points A, B, C, D, E, F, and G.

Delaunay triangulations also match with Voronoi diagrams in the sixth section of each chapter. In Figure 4.18, the solid lines form the Voronoi diagram for
the seven points A, B, C, D, E, F, and G. The dotted lines form the Delaunay triangulation for the same seven points. The points at the intersections of the Voronoi
diagram are the centers of the circles through the vertices of the triangles in the
Delaunay triangulation. For instance, the point furthest to the left in Figure 4.18
is equidistant from the points A, E, and F. See [11, 79–86] for more on Delaunay
triangulations.

Problem 3.4 (c). (repeated) Explore whether polygons with n vertices (including the vertices of the holes) and h interior holes made of polygons always have

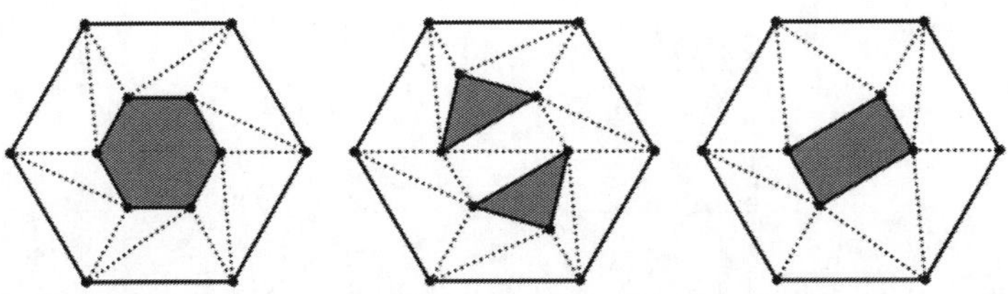

Figure 4.19. Triangulations of polygons with holes. (a) [left] $n = 12$,
$h = 1$, $T = 12$. (b) [center] $n = 12$, $h = 2$, $T = 14$. (c) [right] $n = 10$,
$h = 1$, $T = 10$.

the same number of triangles in a triangulation. If so, find a formula for the number of triangles in a triangulation in terms of n and h.

Recall from Theorem 2.3 that a polygon with n vertices (and no holes) has $n - 2$ triangles in a triangulation. A few well chosen examples with holes, as in Figure 4.19, suggest the formula $T = n + 2h - 2$ in Theorem 4.3 for the number T of triangles, although not a proof.

Theorem 4.3. A polygon with h polygonal holes and a total of n vertices (including those for the holes) has $n + 2h - 2$ triangles in any triangulation.

Idea for proof. For each hole we delete one edge of the hole and one edge of the outside polygon and connect the outside with the hole to create a new polygon without any holes. (See Figure 4.20.) The number of vertices remains the same, but we lose two triangles for each hole we eliminate. The new polygon has $n - 2$ triangles, and so the original polygon with h holes has $n + 2h - 2$ triangles in a triangulation. $\square$

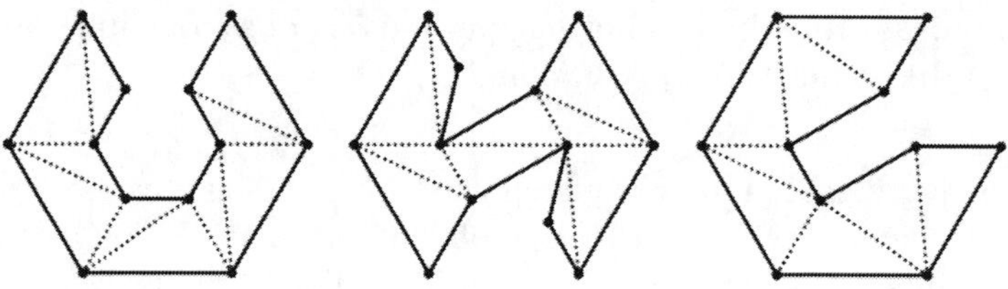

Figure 4.20. Modifying polygons to eliminate holes.

For a complete proof, we'd need to show that this idea can be modified to always work. For instance, we have a problem when there are more holes than edges of the outside polygon. Figure 4.21 indicates that we can delete edges from two holes and connect them to eliminate a hole. A more formal method involves induction.

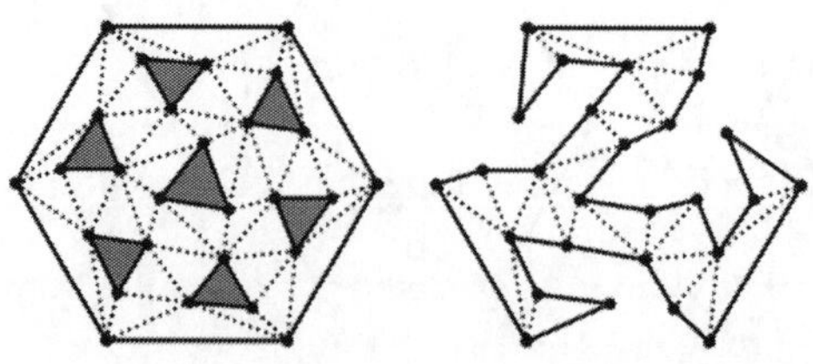

Figure 4.21. Modifying a polygon with multiple holes. (a) [left] $n = 27$, $h = 7, T = 39$. (b) [right] $n = 27, h = 0, T = 25$.

Figure 4.22. Polygons with only one triangulation.

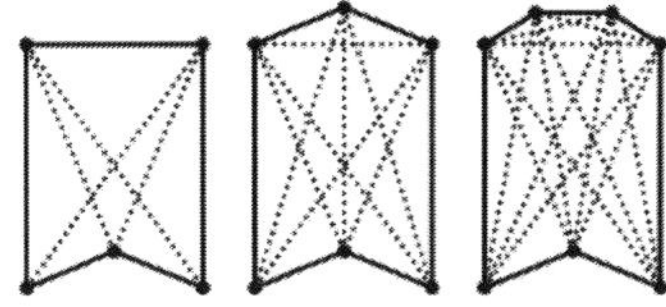

Figure 4.23. Polygons with one fewer diagonal than a convex polygon.

Problem 3.4 (d). (repeated) What is the largest and smallest number of triangulations of a nonconvex polygon with n vertices?

Figure 4.22 indicates that the smallest number of triangulations of a polygon is always one since all of the diagonals must come from the same vertex. For the largest number of triangulations of a nonconvex polygon, intuitively we need as many diagonals as possible. Figure 4.23 illustrates that a nonconvex polygon with n vertices can have up to $\frac{n(n-3)}{2} - 1$ diagonals in a convex polygon, one fewer than the number of diagonals in a convex polygon. By enumeration, in Figure 4.23 there are 3 triangulations in the pentagon, 9 triangulations for the hexagon, and (by exhaustion!) 28 triangulations for the heptagon. By themselves, these numbers don't suggest a pattern and I doubt anyone wants to count the case for an octagon. There has to be a better way to find a general formula for the largest number of triangulations of a nonconvex polygon.

Recall for convex polygons there are 5 for the pentagon, 14 for the hexagon, and 42 for the heptagon. For various numbers of vertices, Table 4.6 gives the number of triangulations in a convex polygon and in a nonconvex polygon with the most triangulations. Let's hope this suggests a pattern.

Table 4.6. Triangulations for convex and "nearly convex" polygons.

n	4	5	6	7
convex	2	5	14	42
nonconvex	1	3	9	28

Each value on the bottom row is the difference of the value above it minus the previous convex triangulation: $3 = 5 - 2$, $9 = 14 - 5$, and $28 = 42 - 14$. Why is that?

Theorem 4.4. (Rogers, 2015) The largest number of triangulations a nonconvex polygon with n vertices can have is $T_n - T_{n-1}$, where T_n is the number of triangulations of a convex n-gon.

Proof. Suppose a nonconvex polygon with n vertices has $\frac{n(n-3)}{2} - 1$ diagonals, one less than the number a convex n-gon has. As in Figure 4.24 there must be exactly two vertices, labelled P_{n-1} and P_1, for which the segment $\overline{P_{n-1}P_1}$ joining them is outside the polygon and one vertex, P_n adjacent to both of them. Reflect P_n and the edges connecting it to P_{n-1} and P_1 over $\overline{P_{n-1}P_1}$ to get the dashed segments $\overline{P_1Q}$ and $\overline{P_{n-1}Q}$. By replacing P_n with Q we make a convex n-gon, which has T_n triangulations. How many of them use the segment $\overline{P_{n-1}P_1}$ as a diagonal? On one side of $\overline{P_{n-1}P_1}$ is $\triangle P_1 Q P_{n-1}$. On the other side is a convex $(n-1)$-gon with T_{n-1} triangulations. Each of these T_{n-1} triangulations together with $\overline{P_{n-1}P_1}$ gives a triangulation of the convex n-gon. Every other triangulation of the convex n-gon avoids $\overline{P_{n-1}P_1}$ and so is a triangulation of the original nonconvex polygon. Hence the original nonconvex polygon has $T_n - T_{n-1}$ triangulations. $\square$

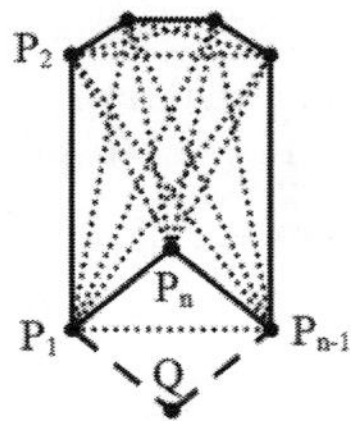

Figure 4.24. Modifying a nonconvex polygon to obtain a convex one.

A former student of the author's, Sam Rogers, found interesting partial results for the range of triangulations possible for nonconvex polygons. However, there appear to be too many cases as n increases to give a complete general answer.

Problem 3.4 (e). (repeated) Look for other pairs of polyhedra besides bipyramids and prisms that switch the number of vertices and faces. Given a polyhedron, can we find another polyhedron with this relationship?

The number of vertices of an n-gonal prism $(2n)$ equals the number of faces of an n-gonal bipyramid. Similarly, the number of faces of that prism $(n+2)$ equals the number of vertices of the bipyramid. These are examples of dual polyhedra.

Definition. Two polyhedra are *duals* provided the vertices of one correspond to the faces of the other and two vertices of one are adjacent exactly when the corresponding faces of the other are adjacent.

As a consequence of the definition, the numbers of vertices and faces switch between dual polyhedra and they have the same number of edges. This fits with Euler's formula (Theorem 3.3) $V - E + F = 2$ for convex polyhedra.

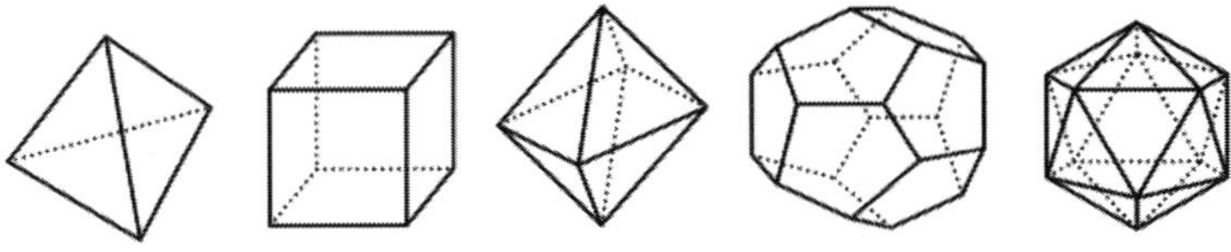

Figure 3.32. (repeated) The five regular polyhedra.

Four of the five regular polyhedra come in dual pairs: the cube and octahedron are duals, as are the dodecahedron and the icosahedron. (See Figure 3.32, reproduced here.) For many convex polyhedra, its dual has the same symmetries as the original polyhedron. Consider the cube and the octahedron. A 90° rotation around the centers of opposite faces of a cube corresponds to a 90° rotation around opposite vertices of an octahedron. Similarly, a 120° rotation around opposite vertices of a cube matches a 120° rotation around centers of opposite faces of an octahedron.

Pyramids are *self-dual*, which Figure 4.25 illustrates, where the shaded dual pyramid fits upside down in the original pyramid.

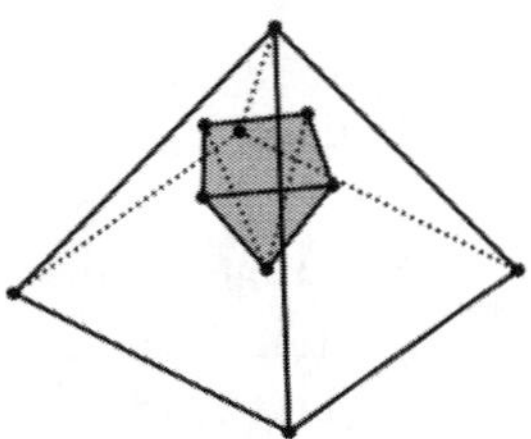

Figure 4.25. A pyramid and its dual.

Exercise 4.3. Look for other self-dual polyhedra besides pyramids. What properties do they have?

Dual polyhedron have delighted mathematicians for centuries. See [35] for much more on this topic.

Exercise 4.4. Explore whether the polyhedron with a hole in Figure 4.26 has a dual and, if so, whether their symmetries match and whether it is self-dual. Generalize.

Not every polyhedron has a dual that is an actual polyhedron, provided we allow polyhedra with holes. The Császár polyhedron has fourteen triangular faces with just seven vertices and twenty-one edges. It also has a hole and no diagonals. (See [10, 37–39] for instructions on making one. A two-dimensional drawing doesn't suffice for visualizing it.) Each vertex has six triangles adjacent to it.

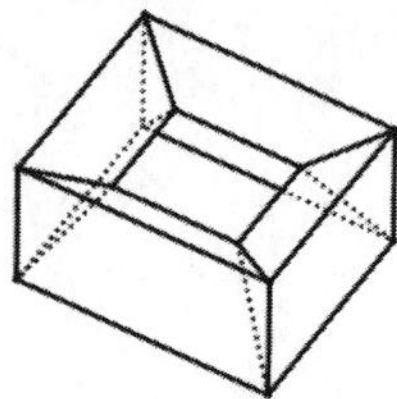

Figure 4.26. A polyhedron with a hole.

The dual therefore would be made from seven hexagons and each hexagon would need to share an edge with each of the other hexagons. It is impossible for each of these hexagons to lie flat and yet share an edge with each of the other hexagons.

Ákos Császár (1924–2017), after whom the Császár polyhedron is named, was a distinguished Hungarian mathematician who survived being in a Nazi concentration camp late in World War II. He constructed the polyhedron and proved its properties early in his career, in 1949.

Problem 3.4 (f). (repeated) Investigate equalities or inequalities linking V and E or E and F. Investigate what values of V, E, and F are possible.

The polyhedron with the fewest number of vertices, edges, and faces is a tetrahedron, with $V = 4$, $E = 6$, and $F = 4$. For any polyhedron each edge separates two faces (and has two vertices at its ends). Also every face has to have at least three edges. Thus, $3F \leq 2E$. (For instance, a cube with six faces and twelve edges satisfies $3 \cdot 6 \leq 2 \cdot 12$.) Similarly, every vertex has at least three edges adjacent to it in order for the polyhedron to be three-dimensional. So, $3V \leq 2E$. We can use these inequalities to replace E in $V - E + F = 2$. With $3F \leq 2E$, we have $\frac{3}{2}F \leq E$ and so $-\frac{3}{2}F \geq -E$. Thus $V - \frac{3}{2}F + F \geq 2$ or $V - 2 \geq \frac{1}{2}F$ or $2V - 4 \geq F$. Similarly, from $3V \leq 2E$ we have $2F - 4 \geq V$, or by solving for F, $F \geq \frac{1}{2}V + 2$.

Figure 4.27 graphs the lines $F = 2V - 4$ and $F = \frac{1}{2}V + 2$ and shades in the region satisfying both inequalities, which represents values of V and F of possible polyhedra. In addition, the dashed lines indicate the number of edges there would be on a polyhedron with the corresponding numbers for V and F. Surprisingly, the value $E = 7$ is impossible, although every other value of E greater than 5 is possible.

Do all the other values of V, E, and F in the shaded region correspond to actual polyhedra? Yes. We can build such polyhedra recursively starting from pyramids, which correspond to points on the diagonal $F = V$. From the slopes of the two lines of Figure 4.27, we need to be able to add either one face and two vertices (for the slope of $\frac{1}{2}$) or two faces and one vertex (for the slope of 2). Figure 4.28 illustrates how to do this to a given polyhedron as long as it has a triangle

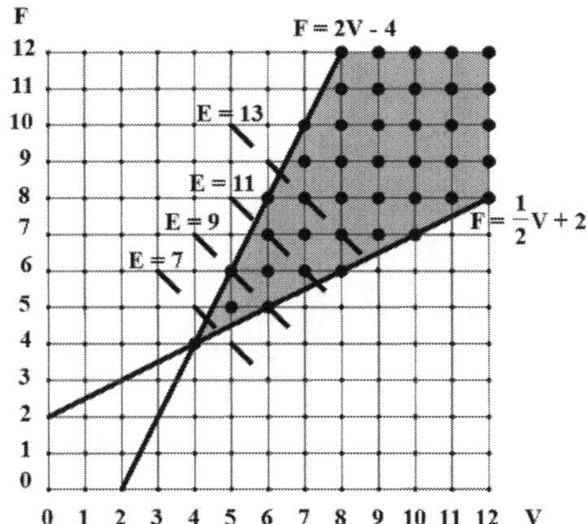

Figure 4.27. The shaded region represents possible values of V and F for polyhedra. The dashed lines give values of E for these V and F.

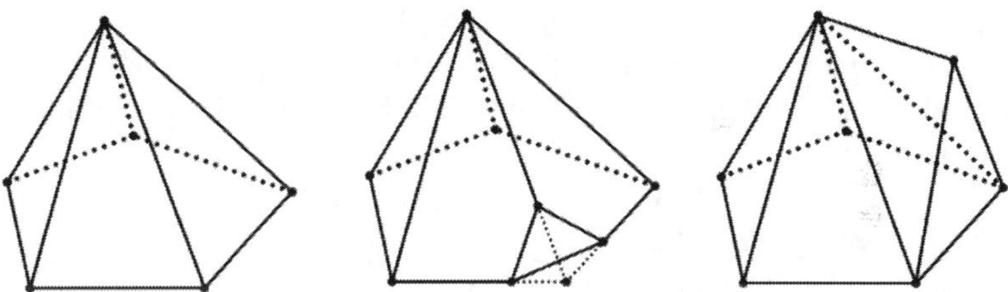

Figure 4.28. Modifying a polyhedron to add vertices and faces. (a) Original polyhedron, $V = 6$, $E = 10$, $F = 6$. (b) Truncated version, $V = 8$, $E = 13$, $F = 7$. (c) Stellated version, $V = 7$, $E = 13$, $F = 8$.

for a face and some vertex with three faces adjacent. In each case, we add three edges. Figure 4.28 (b) truncates a vertex of the original polyhedron of Figure 4.28 (a) by cutting off the tip and replacing it with a triangle. Figure 4.28 (c) stellates a triangular face of Figure 4.28 (a), replacing the face with a pyramid with three faces and adding one vertex.

Now that we know every possible combination of V, E, and F has an actual polyhedron, we could ask can there be more than one type? That is, are there different numbers of triangles, quadrilaterals, etc. that can give the same values of V, E, and F? Yes. The smallest number of edges where this occurs is $E = 10$

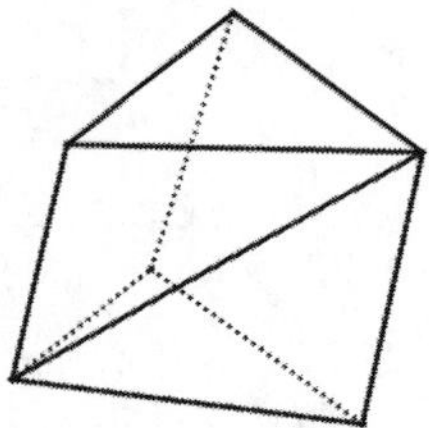

Figure 4.29. A polyhedron with $V = 6$, $E = 10$, and $F = 6$.

with both $V = 6$ and $F = 6$. Figure 4.29 illustrates a bent triangular prism with one of its quadrilateral faces bent in or out to make two triangles, for a total of two quadrilaterals and four triangles. We already know a pentagonal pyramid satisfies these values.

Problem 3.4 (g). (repeated) Investigate $V - E + F$ for nonconvex polyhedra to determine conditions needed for a polyhedron to satisfy Euler's formula. For those, if any, that don't satisfy Euler's formula, look for a more general formula.

Many nonconvex polyhedra still satisfy Euler's formula, $V - E + F = 2$, provided they don't have holes in them. Henri Poincaré (1854–1912) generalized and proved Euler's formula for polyhedra with holes and also for n-dimensional generalizations of polyhedra, called *polytopes*. We give without proof Poincaré's theorem for polyhedra with holes, illustrated in Figure 4.30. See [6, 31–32] for more on polyhedra with holes.

Theorem 4.5 (Poincaré's theorem). (Poincaré, 1899) If a polyhedron has H holes, V vertices, E edges, and F faces, then $V - E + F = 2 - 2H$.

Proof. See [4, 232–233]. $\square$

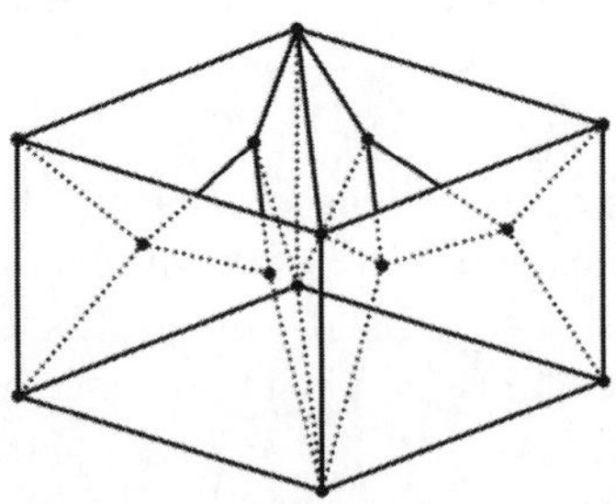

Figure 4.30. A polyhedron with two holes, $V = 14$, $E = 32$, $F = 16$, and $V - E + F = -2 = 2 - 2H$.

Exercise 4.5. Use Poincaré's theorem (Theorem 4.5) to verify that the Császár polyhedron with $V = 7$, $E = 21$, and $F = 14$ must have one hole. How can you determine that all of its faces must be triangles, that there are no diagonals, and that each vertex has six edges and so six triangles adjacent to it?

Poincaré's higher-dimensional generalization is important in topology and other areas of mathematics. As a temptation, let's look at a two-dimensional rendering of a *hypercube* in Figure 4.31 and count its vertices, edges, faces, and cubes to illustrate Poincaré's formula for polytopes in four dimensions without holes. (A hypercube is the four-dimensional analog of a cube.) For a convex polytope in four dimensions, Poincaré proved that $V - E + F - C = 0$, where C is the number of three-dimensional cells. For a hypercube, the cells are ordinary cubes. We have $V = 16 = 2^4$. Each vertex is adjacent to four edges, giving a total of $E = \frac{16 \cdot 4}{2} = 32$ since each edge has two ends. Any two edges at a vertex determine a square face, so there are $\binom{4}{2} = 6$ faces at each vertex. This gives $F = \frac{16 \cdot 6}{4} = 24$ since each face has four corners. Finally, any three of the four edges at a vertex determine a cube. So, there are four cubes at each vertex and $C = \frac{16 \cdot 4}{8} = 8$ since a cube has eight vertices. Then $V - E + F - C = 16 - 32 + 24 - 8 = 0$, what Poincaré's formula predicts. The hypercube is one of the regular polytopes—higher-dimensional analogues of the regular polyhedra. Its four-dimensional dual is also regular and has $V = 8$, $E = 24$, $F = 32$, and $C = 16$. Its faces are all equilateral triangles and its cells are regular octahedra. (We'll encounter this dual in our answer to Problem 3.5 (c).) See [22] for more on this fascinating area.

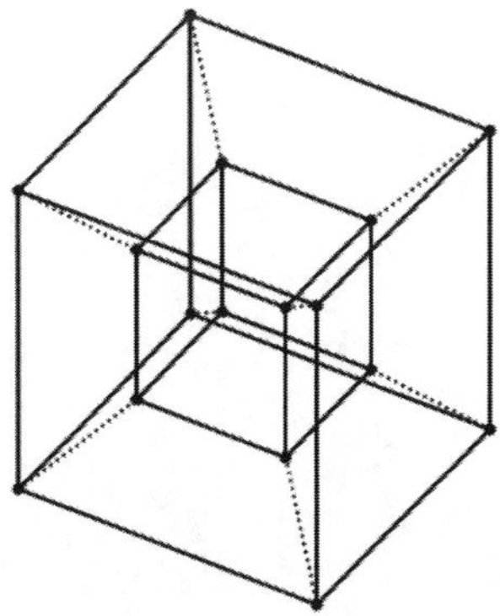

Figure 4.31. A two-dimensional projection of a hypercube.

Problem 3.4 (h). (repeated) How can we modify Descartes' formula for non-convex polyhedra?

Poincaré's theorem for polyhedra with holes (Theorem 4.5) enables us to modify Descartes' formula for the angle sum of a polyhedron with holes. Recall in the proof of Descartes' formula we rearranged $V-E+F = 2$ to get $V-2 = E-F$. If there are H holes, we rearrange $V - E + F = 2 - 2H$ to get $V - 2 + 2H = E - F$. When we multiply the equation by $360°$, the right side, $360°(E - F)$ still counts the angle sum as before. So a polyhedron with V vertices and H holes has an angle sum of $360°(V - 2 + 2H)$.

Examples. The polyhedron with $H = 1$ hole in Figure 3.24 has $V = 12, E = 24$ and $F = 12$. The twelve quadrilateral faces have an angle sum of $360°(12) = 360°(12 - 2 + 2 \cdot 1) = 360°(V - 2 + 2H)$. The polyhedron in Figure 4.30 has $V = 14, E = 32, F = 16$, and $H = 2$. Again, all faces are quadrilaterals and $360°(V - 2 + 2H) = 360°(16)$.

Problem 3.4 (i). (repeated) Investigate tetrahedralizing nonconvex polyhedra.

Many nonconvex polyhedra can be tetrahedralized. However, in 1911, Nels Lennes (1874–1951) published the surprising example of a nonconvex polyhedron that can't be tetrahedralized. The simplest example has just six vertices. Think of a squat triangular prism (Figure 4.32) and twist the top relative to the bottom just enough so that the rectangular faces become two triangles bent in and sharing a common edge, as in Figure 4.33. (This represents even more twisting than the polyhedron of Figure 4.29.) The resulting polyhedron made of eight triangles has no surface diagonals and no interior diagonals. Each vertex has four edges and one "exterior diagonal" to the remaining vertex, such as $\overline{BF}$ in Figure 4.33. Any four of the vertices includes a pair with an exterior diagonal and so is not a tetrahedron entirely contained inside the polyhedron.

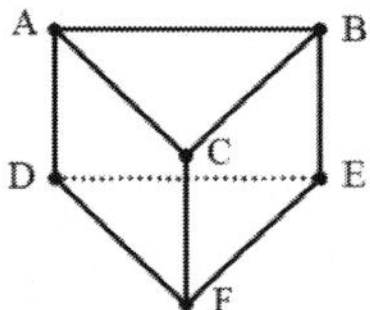
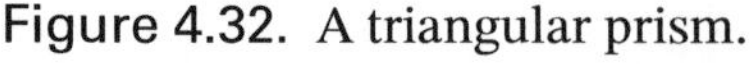

Figure 4.32. A triangular prism.

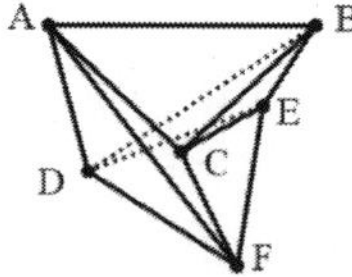

Figure 4.33. A twisted prism.

4.3 Distances and Points

Problem 3.5 (a). (repeated) Find a formula for the number of points in a hexagonal arrangement of points with k circular layers. Find a formula for the

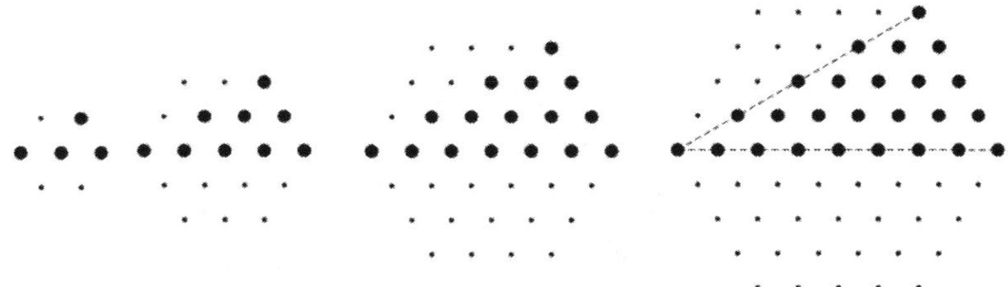

Figure 4.34. Hexagonal arrangements of points with one, two, three, and four circular layers. The larger dots help count the number of different distances.

number of distances determined by k circular layers, ignoring the duplicates using the law of cosines.

Let's start with smaller hexagonal arrangements, as in Figure 4.34, and look for a pattern from the values in Table 4.7.

Table 4.7. Number of points and distances for k layers of a hexagonal arrangement of points. (* ignoring the law of cosines)

layers	1	2	3	4
points	7	19	37	61
distances	3	8	15	24*

The distances (ignoring the reduction from using the law of cosines) are one less than square numbers: $(k + 1)^2 - 1 = k^2 + 2k$. Difference equations from Section 3.1 enable us to find the formula $3k^2 + 3k + 1$ for the number of points in k layers. Not surprisingly, the OEIS has this sequence, $A003215$.

Problem 3.5 (b). (repeated) Find an arrangement in three dimensions of more than twenty points that you think has a minimum number of nonzero distances. Try to describe a general arrangement in three dimensions to minimize the number of distances.

The hexagonal arrangement in two dimensions seems a good place to start for three dimensions. We might first think of stacking layers of this directly on top of one another. However, there is a somewhat better way, where the layers nestle in between. This arrangement connects with two other geometric properties. First, the cuboctahedron, shown in Figure 4.35, has the singular property that it is the only polyhedron all of whose vertices are the same distance from its center as they are from their neighbors. Notice that some slices through the cuboctahedron intersect it in a regular hexagon. We can use one of the regular

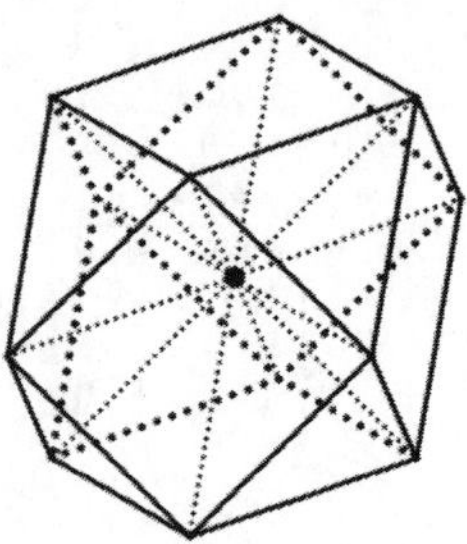

Figure 4.35. A cuboctahedron with its center connected to its vertices.

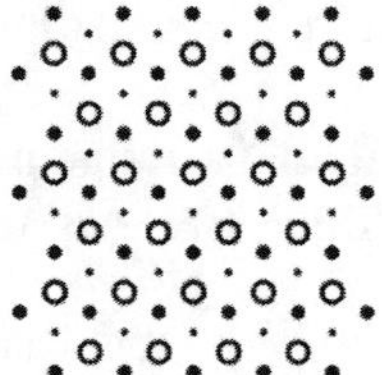

Figure 4.36. Nestled layers of hexagonal arrangements of points in three dimensions.

hexagons as the start of a hexagonal arrangement of points in a plane. In addition, there are equilateral triangles above and below the hexagon that can be the seeds for hexagonal arrangements of points. Figure 4.36 illustrates three layers of such hexagonal arrangements with the small dots being the lowest layer, the bigger dots the next layer and the circles the higher layer. We can stack more layers directly on top of these in the same order. These match the centers of spheres in the closest packing of spheres in three dimensions—the subject of Problem 3.5 (e).

In 1946, Paul Erdős (1913–1996) posed and partially solved the problem of finding the minimum number of distances for points in any number of dimensions. Let $g_d(n)$ be the minimum number of distinct distances n points can determine in d-dimensional Euclidean space. Erdős proved that $\sqrt{n - \frac{3}{4}} - \frac{1}{2} \leq g_2(n) \leq \frac{cn}{\log(n)}$, for some constant number c. For our hexagonal arrangement of $3n^2 + 3n + 1$ points, we have at most $n^2 + 2n$ distinct distances, a bit more than $\frac{1}{3}$ of the number of points, suggesting that the constant c might be about $\frac{1}{3}$. The term $\log(n)$ in the denominator comes from the reductions we get from duplications using the law of cosines.

For $n < m$ and $d < f$, we have $g_f(n) \leq g_d(n) \leq g_d(m)$. That is, a higher dimension can lower the minimum number of distances for a given number of points, as can having fewer points. Table 3.6 illustrated these inequalities. Erdős showed that for large values of n and for a given dimension $d > 2$, there is a constant c so that $g_d(n) \leq cn^{2/d}$. For $d = 3$, we have $cn^{2/3}$. For instance, when $n = 1000 = 10^3$, we have $100c$ and when $n = 1,000,000 = 10^6$, we have $10,000c$. Since 1946 several mathematicians have contributed more refined upper bounds for different dimensions.

Paul Erdős was a prolific mathematician, coauthoring more papers than any other mathematician and rivalling Leonhard Euler for the most mathematical publications of all time. He is also noted for posing challenging open problems along with prize money for solutions. However, the honor of solving an Erdős problem was so great that many mathematicians who have solved one of these problems have framed the check, rather than cash it. Between the World Wars, as a Jew, Erdős was forced to leave his home country of Hungary. He held a variety of short-term university positions, but he was famously eccentric and unsuited as a professor. After 1963, he had no home and no job for the rest of his life. However, mathematicians were eager to invite him to their university to get the chance to collaborate with him. He used to sum up his nomad life style as "another roof, another proof." (For a biography of Erdős, see [19].)

Problem 3.5 (c). (repeated) Generalize taxicab distance to four dimensions and repeat Problem 2.5 (d) for four dimensions.

The taxicab metric readily extends to four dimensions:

$$d_t((p,q,r,s),(t,u,v,w)) = |t - p| + |u - q| + |v - r| + |w - s| \, .$$

Before finding arrangements for three, four, and five distances, let's start with one distance, 2, as we did in three dimensions. The eight points $(\pm 1, 0, 0, 0)$, $(0, \pm 1, 0, 0)$, $(0, 0, \pm 1, 0)$, and $(0, 0, 0, \pm 1)$ are at a distance of two from each other. They are the vertices of the regular polytope called variously a *cross polytope* or a *hyperoctahedron* that we mentioned as the dual of the hypercube in our discussion of Poincaré's generalization of Euler's formula. We can imitate the approach from the response to Problem 2.5 (c) to find points with two distances, 2 and 4. Since a drawing is impractical, we provide a listing of the 33 possible points, where the sum of the absolute values of the coordinates of a point is 2 or 0.

$(\pm 2, 0, 0, 0), (0, \pm 2, 0, 0), (0, 0, \pm 2, 0), (0, 0, 0, \pm 2), (0, 0, 0, 0)$
$(\pm 1, 1, 0, 0), (\pm 1, 0, 1, 0), (\pm 1, 0, 0, 1), (0, \pm 1, 1, 0), (0, \pm 1, 0, 1), (0, 0, \pm 1, 1),$
$(\pm 1, -1, 0, 0), (\pm 1, 0, -1, 0), (\pm 1, 0, 0, -1), (0, \pm 1, -1, 0), (0, \pm 1, 0, -1),$
and $(0, 0, \pm 1, -1)$.

For the three distances of 2, 4, and 6, we use all arrangements of points so that the sum of the absolute values of their coordinates is 3 or 1. These include ± 3 with three 0s, all arrangements of ± 2, ± 1 and two 0s, all arrangements of three ± 1s and a 0, and the eight points with ± 1 and three 0s. There are 96 such points. For the four distances 2, 4, 6, and 8, the sums of the absolute values of the coordinates of points can be 4, 2, or 0. There are 225 such points. In our friend OEIS, this sequence has number $A014820$.

Exercise 4.6. Use the reasoning in the previous paragraph to conjecture the maximum number of points in four-dimensional taxicab geometry for five and six distances.

The values for three dimensions we found in the answer to Problem 2.5 (e) form the sequence $A005900$, called the octahedral numbers. These are also the number of spheres of radius 1 packed in the shape of an octahedron.

Problem 3.5 (d). (repeated) Find the dimensions of the smallest squares that contain n nonoverlapping unit circles, for $n = 3, 5, 8,$ and 9. Find the corresponding density. (The densest packings for these and some other values of n have helpful symmetry.)

Refer to Figure 4.37, which illustrates the smallest squares for the required values. For three circles, make them mutually tangent and one tangent to two sides of the square. To find the density let's measure angles: $\angle ABC$ is 60°, $\angle CBD$ is 30°, $\angle DBE$ is 45°, and so $\angle CBE$ is 15°. Since $\overline{BC}$ has length 2, $\overline{BE}$ has length $2\cos(15°) = \frac{\sqrt{6}+\sqrt{2}}{2} \approx 1.9319$ and the square has side $2 + 2\cos(15°) \approx 3.9319$. This gives a density of $\frac{3\pi}{(2+2\cos(15°))^2} \approx 0.6096$.

With five circles, the length of the side of the square is $2+2\sqrt{2}$ and the density is $\frac{5\pi}{(2+2\sqrt{2})^2} \approx 0.6738$. The design for eight circles incorporates the arrangement of three circles in each corner. So, the side of the square is $2 + 4\cos(15°) \approx 5.8637$, giving a density of 0.7310. With nine circles we are back to part of the periodic square array with a density of $\frac{\pi}{4} \approx 0.7854$.

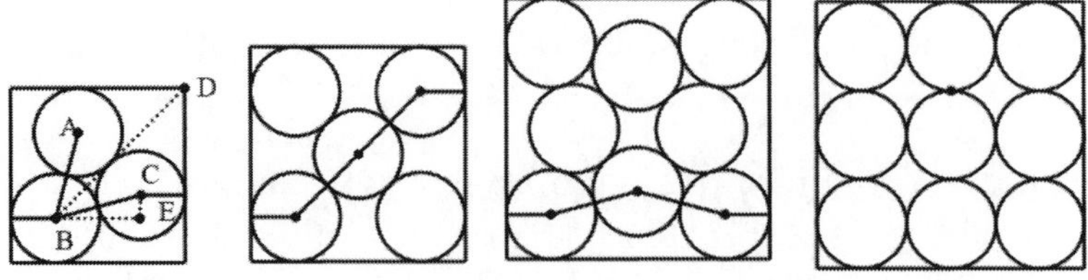

Figure 4.37. Smallest squares with 3, 5, 8, and 9 unit circles.

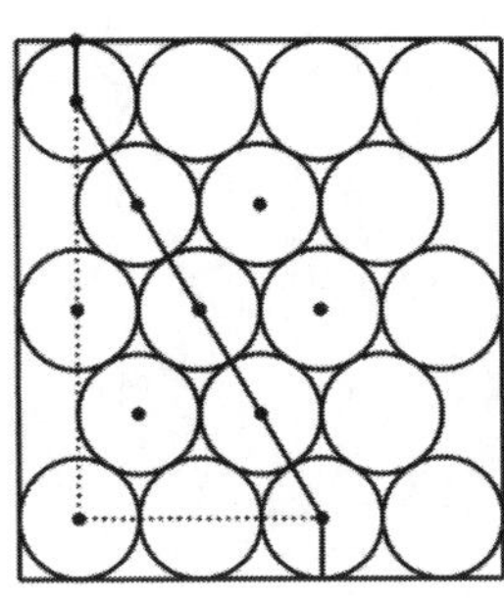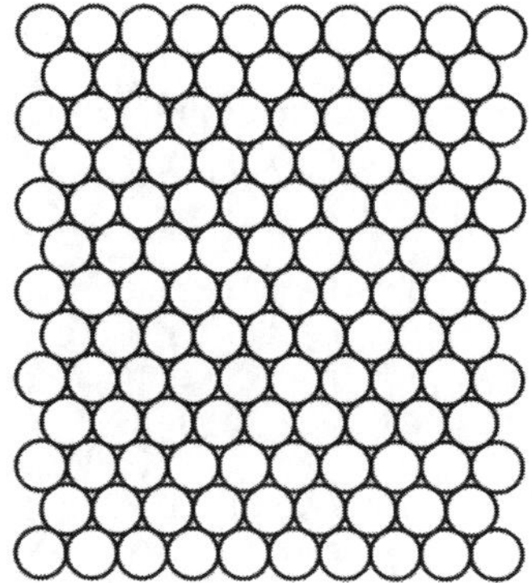

Figure 4.38. Hexagonal arrangements of circles in a square.

It is possible to obtain higher densities using the hexagonal packing density than a square packing with a large enough number of circles. The left of Figure 4.38 illustrates the arrangement of eighteen unit circles in a rectangle with dimensions $8 \times (2 + 4\sqrt{3})$ or approximately 8×8.9282. So this arrangement has a density of $\frac{18\pi}{(2+4\sqrt{3})^2} \approx 0.7094$, still not as much as the density of the square arrangement of circles. The hexagonal array of 105 circles on the right of Figure 4.38 fits in a 20×20 square, giving a density of $\frac{105\pi}{400} \approx 0.8247$. For large numbers of circles, the hexagonal has been proven to be the best.

Problem 3.5 (e). (repeated) Find some periodic packings of spheres and determine their densities.

The easiest packing of spheres to visualize stacks square arrays on top of one another. Each sphere fits inside a cube with side 2, giving a density of $\frac{4\pi/3}{8} \approx 0.5236$. Johannes Kepler (1571–1630) found a denser packing, which he conjectured was the densest possible. It uses layers of hexagonal arrays with each layer nestling on the lower level so that a sphere in one level touches three spheres in the lower level, as in Figure 4.39. This corresponds to the arrangement of points discussed in the response to Problem 3.5 (b). The density of Kepler's arrangement is $\frac{\pi}{3\sqrt{2}} \approx 0.7405$. The centers of the twelve spheres tangent to a given sphere match the vertices of a cuboctahedron, pictured in Figure 4.35. Since some of the faces of a cuboctahedron are squares, some mathematicians after Kepler suspected that one might be able to adjust the spheres around a sphere to squeeze in a thirteenth sphere. Carl Friedrich Gauss (1777–1855) showed in 1831 that Kepler's packing was the densest periodic packing possible. Thomas Hale (1958–) in an exceptionally difficult, computer-aided argument showed in 1998 that no packing, periodic or not, could be more dense. Assisted by others, he provided a fully accepted formal proof of his result in 2014.

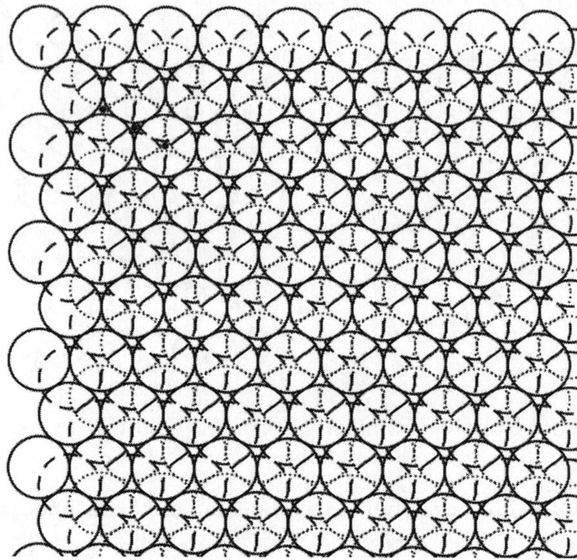

Figure 4.39. Layers of spheres in Kepler's packing. The solid circles represent the lowest level, the dashed ones the middle, and the dotted ones the top.

Sphere packing (or more accurately *hypersphere* packing) in higher dimensions is a long-standing and difficult problem. Mathematicians have found the densest packings for dimensions up to eight. In 2022, Maryna Viazovska (1984–) was awarded the Fields Medal for her proof in 2016 giving the densest packing of identical hyperspheres in eight dimensions, as well as other important contributions in mathematics. (The density is $\frac{\pi^4}{2^4 4!} \approx 0.25$.) The Fields Medal is the most prestigious prize in mathematics. In collaboration with others, she also proved in 2017 the densest hypersphere packing in twenty-four dimensions (with density $\frac{\pi^{12}}{12!} \approx 0.0019$). You may think higher-dimensional problems are purely theoretical musings. However, mathematicians and computer scientists have used the eight and twenty-four–dimensional arrangements to craft efficient *error correcting codes*. (For more, see [33].)

When a computer sends information there is a small chance of a random error in transmission—a sent 0 gets received as 1 or 1 as 0. The receiving computer needs a way to recognize an error and correct it. The key is to build a space around each legitimate string of 0s and 1s so that an error will stand out. This matches with the idea of sphere packing: the center of each sphere is a legitimate "codeword" and every string of symbols differing in one place (or several places for fancier codes) is inside the sphere. If the spheres don't overlap, the computer can recognize an error or errors have happened and find the correct codeword closest to what was received.

Problem 3.5 (f). (repeated) Define a taxicab circle and sphere. Explore circle and sphere packing in taxicab geometry.

In ordinary geometry, a circle and a sphere are the sets of all points a fixed distance (the radius) from a point (the center) in two or three dimensions. We'll use the same definitions for taxicab circles and spheres. However, as we see for circles of radius 1 and 2 in Figure 4.40, a *taxicab circle* is a square. A taxicab sphere, as in Figure 4.41 is a regular octahedron. The equation for a taxicab circle of radius 1 centered at the origin is $|x| + |y| = 1$. Similarly, a *taxicab sphere* has equation $|x| + |y| + |z| = 1$.

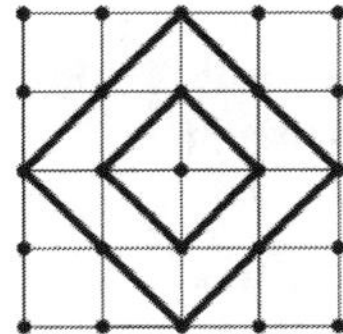

Figure 4.40. Taxicab circles.

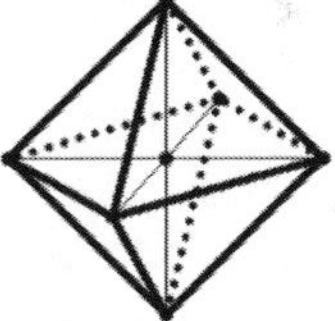

Figure 4.41. A taxicab sphere.

Circle packing in taxicab geometry is easy. As Figure 4.42 indicates, we can completely fill the plane with taxicab circles, giving a density of 1. Taxicab spheres in space are harder to visualize. So, let's first think about coordinates for taxicab circles of radius 1 and use them to help us with spheres. If the center of one of the taxicab circles is at $(0, 0)$, then its vertices are at $(\pm 1, 0)$ and $(0, \pm 1)$. Adjacent circles will have centers at $(\pm 1, 1)$ and $(\pm 1, -1)$. Further out are circles

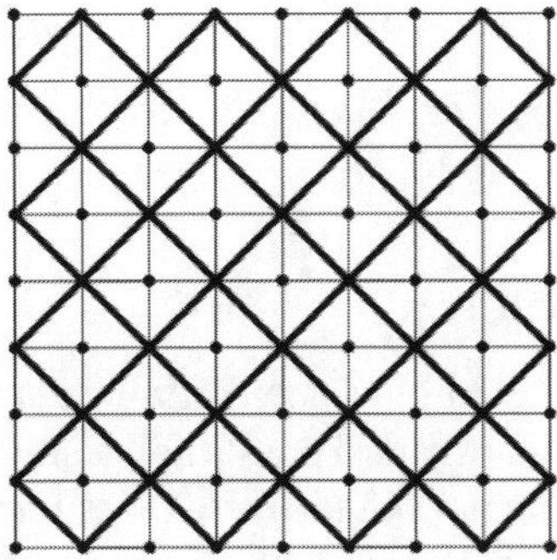

Figure 4.42. Taxicab circles of radius 1 cover the plane.

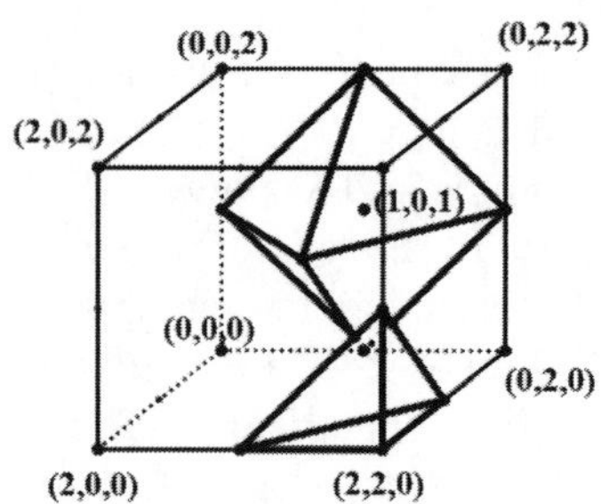

Figure 4.43. Parts of two taxicab spheres in a $2 \times 2 \times 2$ cube.

with centers at $(\pm 2, 0)$ and $(0, \pm 2)$. In general, the centers have coordinates (x, y) with $x + y$ even, and the vertices are at (x, y) with $x + y$ odd.

We can put taxicab spheres of radius 1 with centers (x, y, z), where $x + y + z$ is even, in which case the vertices will have coordinates (x, y, z) with $x + y + z$ odd. However, these spheres don't fill up space. So, we need to determine the density of our sphere packing. Figure 4.43 shows part of two taxicab spheres in a $2 \times 2 \times 2$ cube. The sphere centered at $(1, 0, 1)$ in the middle of the back face of the cube has half of it in this cube. The spheres centered at the centers of the other five faces of the cube also have half of their volume in the cube. The sphere centered at the corner $(2, 2, 0)$ has one eighth of it inside the cube. The same holds for the other seven spheres centered at the other vertices of the cube. We will use Euclidean volume to compute the density. The cube is easy: $2 \times 2 \times 2 = 8$ cubic units. The volume of a pyramid is $\frac{1}{3}(base)(height)$. The pyramids in the corners have a base with area $\frac{1}{2}(1)(1) = \frac{1}{2}$ and a height of 1, for a volume of $\frac{1}{6}$. The square pyramids jutting inward from each face have a square base of $(\sqrt{2})(\sqrt{2}) = 2$ and a height of 1, for a volume of $\frac{2}{3}$. With eight vertices and six faces in a cube, these volumes are together $8(\frac{1}{6}) + 6(\frac{2}{3}) = \frac{16}{3}$. Thus the density of the spherical packing is $\frac{16/3}{8} = \frac{2}{3}$.

4.4 Art Gallery Problems

Problem 3.6 (a). (repeated) Find a polygon whose smallest set of prison guards is the union of a smallest set of guard points and a smallest set of fortress points, but neither the guard points not the fortress points form a set of prison points.

Consider the polygon in Figure 4.44. Its convex hull is an octagon, so we need at least four points in any set of fortress points, such as $\{A, B, C, D\}$. We need at least four guard points because of the four triangles in the four corners.

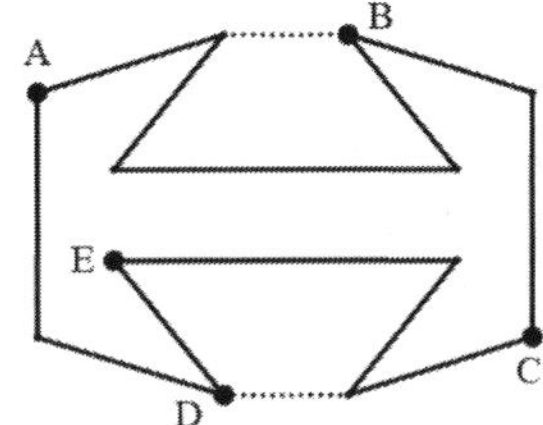

Figure 4.44. A polygon whose minimal set of prison guards is a union of guard points and fortress points.

Furthermore, at least one guard point has to be on one of the interior horizontal segments to see the horizontal strip, such as point E. So one smallest set of guard points is $\{A, B, C, E\}$. Neither of these sets is a set of prison guards, but their union $\{A, B, C, D, E\}$ is.

Problem 3.6 (b). (repeated) Reconsider Problems 1.6, 2.6 (a), and 2.6 (b) where all angles of the polygon are right angles.

Polygons all of whose angles are right angles are called *orthogonal polygons*. In Greek, "ortho" means "right" and "gon" means "angle" or "corner." For ease, we will have the edges be horizontal or vertical. Note that the total number of edges and so vertices will have to be even since the edges have to alternate between being vertical and horizontal. Examples, as in Figure 4.45, with a small number of vertices don't reveal the pattern as well as one might like, as Table 4.8 illustrates. In Figure 4.45, guard points are labelled with a G and fortress and prison points have large dots.

The undulating form of the polygon in Figure 4.46 suggests a way to build a polygon that requires one fourth as many art gallery guards as vertices. It also suggests that we need one additional fortress or prison guard. Several mathematicians have worked on the general problem for orthogonal polygons. As Theorem 4.6 indicates for orthogonal polygons having n vertices, Figure 4.46 gives

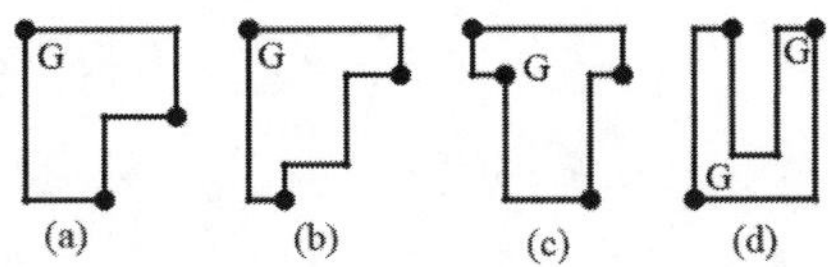

Figure 4.45. Orthogonal polygons with six and eight vertices.

Table 4.8. The number of guard points, fortress points, and prison guards for the orthogonal polygons in Figure 4.45.

Figure 4.45	(a)	(b)	(c)	(d)
guard	1	1	1	2
fortress	3	3	4	3
prison	3	3	4	3

the worst case for the art gallery ($\lfloor \frac{n}{4} \rfloor$) and fortress ($\lceil \frac{n}{4} \rceil + 1$) problems. However, prisons can be more challenging. Theorem 3.9 assures us that the maximum number of prison guards is the sum of the interior and exterior guards, $\lfloor \frac{n}{4} \rfloor + \lceil \frac{n}{4} \rceil + 1 \le \lceil \frac{n}{2} \rceil + 1$. But Joseph O'Rourke lowered that upper bound a bit, although that might not be the best possible upper bound.

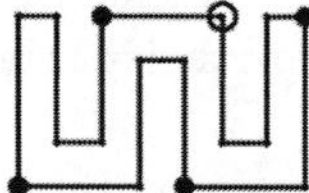

Figure 4.46. A "snake" polygon requiring four art gallery guards and five fortress and prison guards.

Theorem 4.6. (Kahn, Klawe, and Kleitman, 1980; Aggarwal, 1983; O'Rourke, 1983) An orthogonal polygon with n vertices needs at most $\lfloor \frac{n}{4} \rfloor$ guards for the interior, at most $\lceil \frac{n}{4} \rceil + 1$ for the exterior, and at most $\lfloor \frac{7n}{16} \rfloor + 5$ for both.
Proof. See [24, pages 46, 149–150, 157–158]. ☐

Problem 3.6 (c). (repeated) Consider fortress points for convex polyhedra.

Each face of a convex polyhedron needs to have a fortress point. Each vertex is on at least three faces. From this fact we might guess that for a polyhedron with n faces the maximum number of fortress points needed is related to $\frac{n}{3}$. Similarly

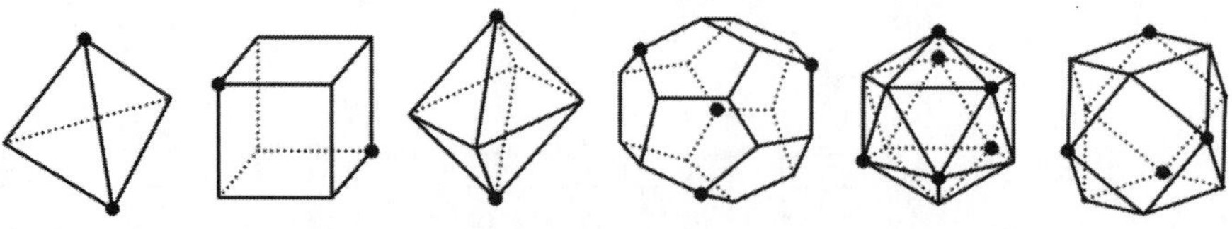

Figure 4.47. The five regular polyhedra and the cuboctahedron with minimal sets of fortress points.

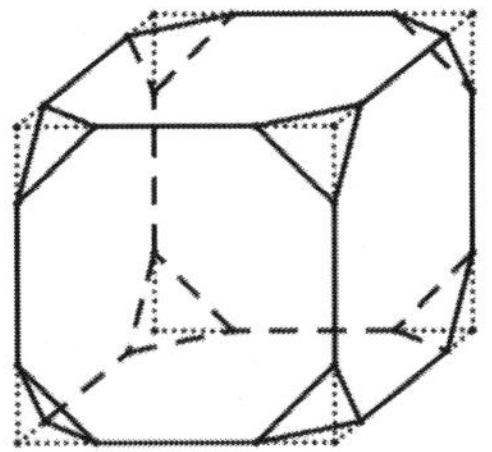

Figure 4.48. A truncated cube.

for polyhedra with n faces altogether and at least k faces at a vertex, we might expect to find $\frac{n}{k}$ of the vertices enough to form a set of fortress points. Figure 4.47 gives the five regular polyhedra and the cuboctahedron as examples with large dots indicating minimal sets of fortress points. All but the icosahedron fit this more general conjecture. The icosahedron has 20 faces with 5 faces per vertex. That suggests a set of four vertices could form a set of fortress points. However, the smallest set needs six vertices. That is still within a conjectured upper bound of $\frac{n}{3} = \frac{20}{3} = 6.66\ldots$.

However there are polyhedra that exceed that bound. Figure 4.48 illustrates a truncated cube made of six octagons and eight triangles. Since no two triangles share a vertex, each one has to have a fortress point at one of its vertices. This forces at least eight fortress points and eight suffice. There are $6 + 8 = 14$ faces altogether, so we would have expected needing at most $\frac{14}{3} = 4.66\ldots$ fortress points, less than the needed 8. This situation illustrates the worst case of Theorem 4.7.

Theorem 4.7. (Grünbaum and O'Rourke, 1983) For a convex polyhedron with n faces and $n \geq 10$, at most $\left\lfloor \frac{2n-4}{3} \right\rfloor$ fortress guards are needed.

Proof. See [24, 257–258]. $\qquad\qquad\square$

Problem 3.6 (d). (repeated) Consider the art gallery problem for polygons with polygonal holes.

From Theorem 4.3, we can triangulate a polygon with holes. So we might think the same formula for the number of guards, $\left\lfloor \frac{n}{3} \right\rfloor$ from Theorem 2.5, will still work. Perhaps surprisingly, an exact formula for this case is not currently known. Figure 4.49 with seven vertices illustrates that we can't always three-color the vertices, which was key to the proof of Theorem 2.5. The number of guards needed is at most $\left\lfloor \frac{n+h}{3} \right\rfloor$, where n is the number of vertices and h is the number of holes ([17, 468]). Figure 4.50 gives an example needing this many guards.

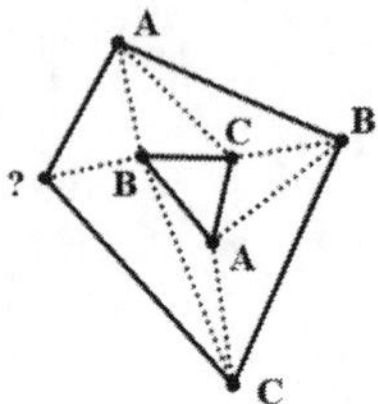

Figure 4.49. A polygon with a hole whose vertices can't be three colored.

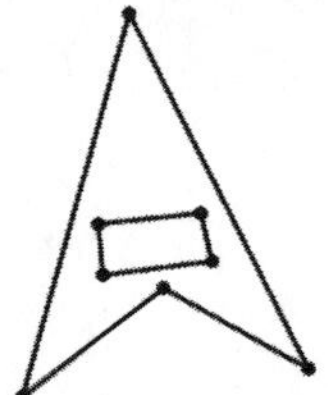

Figure 4.50. A polygon with $n = 8$ and $h = 1$ needing $\left\lfloor \frac{n+h}{3} \right\rfloor = 3$ guards.

Problem 3.6 (e). (repeated) Consider the art gallery problem for nonconvex polyhedra in three dimensions.

The art gallery problem in three dimensions is quite complicated and not much is known. There are polyhedra which have interior points not visible from any vertex. Also, the fact that not every polyhedron has a tetrahedralization makes proofs difficult. See [24, 253–256] for more.

4.5 Geometric Patterns

Problem 3.7 (a). (repeated) Explore conditions on maps that require three colors or more than four colors.

Figure 4.51 gives four maps that seem closely related in pairs, but the ones on the left needs four colors, while the ones on the right needs only three colors. For the map in the lower left, the two regions labelled 2* and 1* seem very similar. Each is surrounded by five cyclically adjacent regions, with four of them adjacent to the surrounded region. Theorem 3.10 doesn't apply to this situation—for it to apply, all five of the cyclically adjacent regions would have to be adjacent to the inside region. Mathematicians have found and proved some criteria guaranteeing that a map needs just three colors, but many maps don't fit these criteria or the ones forcing four colors. This means that currently one has to color these other maps—or for larger maps, have a computer search for a minimal coloring.

In 1976, Kenneth Appel and Wolfgang Haken resolved the four-color conjecture. They published a proof of the four-color map theorem—that no map on the plane ever needs more than four colors. It was the first instance of a proof that required a computer program to check the thousands of complicated cases. At the time, many mathematicians worried about a proof that couldn't be checked by a person or even a team of people. Since then, two other independent approaches, still requiring computers, have also proven this result, satisfying almost

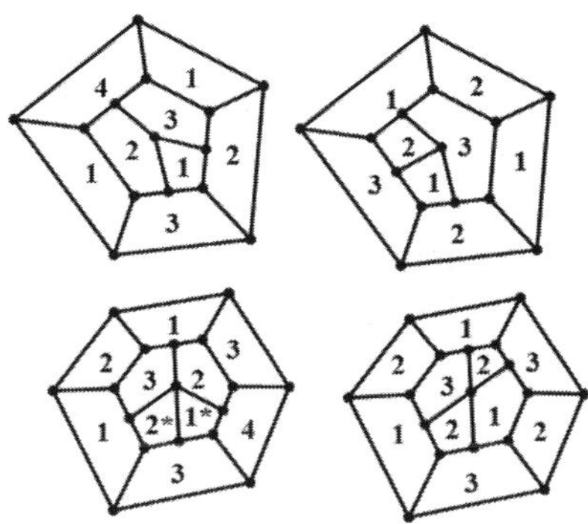

Figure 4.51. Maps requiring three or four colors.

all doubters. The efforts of many mathematicians led up to reducing this complicated problem to a situation where a computer could check the many cases.

Using a computer to determine whether a large, complicated map can be colored with just three colors turns out to be a very difficult programming problem. In fact, L. Stoeckmeyer [32] proved in 1973 that deciding whether a map needs three or four colors qualifies as *NP-complete*—a category indicating it is equally as hard as a large family of really hard problems. There is a million dollar prize for anyone who can find an efficient algorithm to solve any one of these problems or prove that no such efficient algorithm exists. (Most computer scientists in this area believe that no such algorithm exists since people have been searching for such algorithms for decades.) If such an algorithm is found for one of them, it can be converted into an efficient algorithm for any of the others. For more on the million dollar prize for solving any of the seven "millennium problems," see [12]. Let's take a detour to explore the idea of NP-complete.

Computers solve complicated problems using explicit instructions in an algorithm. Computer scientists care greatly about the speed of such algorithms and describe the speed of an algorithm as a function of the size of the inputs. For instance, we might want a computer to sort a list of n names alphabetically. It should seem clear that sorting $n = 1000$ names takes longer than sorting $n = 100$ names. The algorithm has to look at each of the n entries and decide where it belongs in the current list. Just looking at each name already requires n steps. A slow algorithm might take the next word and compare it to each of the already sorted ones. So the second entry gets compared to one name, the third to two names and so on. If we add these up, we get a triangular number, $\frac{n(n-1)}{2} = \frac{1}{2}n^2 - \frac{1}{2}n$. Thus the time to run this program goes up approximately as a multiple of n^2. A faster way is to compare each new word first with the middle word in the current list to decide whether it is before or after it. If it goes before it, then compare it with the word one fourth of the way through and otherwise with the word three fourths of the way through. We continue splitting the remaining possibilities in half. This cuts down on the number of comparisons to a multiple of $n \log(n)$. Computer scientists use the notation $O(n \log(n))$ for an algorithm that can finish a task with n inputs for large values of n in a time that is a multiple of $n \log(n)$. The slower algorithm we described earlier has a running time of $O(n^2)$. Table 4.9 gives some comparisons of how quickly the running time grows for selected values of n. Already with $n = 100$, the difference in computing time between an algorithm with running time $O(n^2)$ and one with $O(n \log(n))$ running time is enormous. The best known algorithm for triangulating a polygon has running time $O(n \log(\log(n))$, even faster than the best sorting algorithm. All of the entries in Table 4.9 are called *polynomials*. That is, the amount of steps or running times grows like some power of the number of inputs.

Table 4.9. The growth of running time for different types of algorithms.

n	100	1,000	10,000
n^3	$1,000,000$	$1,000,000,000 = 10^9$	10^{12}
n^2	$10,000$	$1,000,000 = 10^6$	10^8
$n \log(n)$	200	3,000	40,000
$n \log(\log(n))$	30	477	6,020

There are some algorithms that grow much faster than polynomially—exponentially, for instance. Such algorithms quickly become infeasible to run for even modestly large inputs. For instance, if the running time doubled for each additional input, for $n = 100$ in Table 4.9, we'd have $2^{100} \approx 1.27 \times 10^{30}$. The acronym NP stands for "nondeterministic polynomial." The best known deterministic algorithm for any NP-complete problem, such as determining whether a map is three-colorable, is not polynomial for any power of n. However, if someone offers a solution to one of these problems, a computer can check whether the proposed solution is correct in polynomial time. For many of these problems, there are algorithms involving randomness (nondeterministic) that can give good approximations in polynomial time. For more on the idea of NP-complete, see [15, Chapters 3 and 4].

Problem 3.7 (b). (repeated) The definition of a frieze pattern requires that the design repeat at set intervals, which makes the pattern discrete. Find examples continuous frieze patterns with different symmetry types.

There are only two (rather boring) symmetry types for *continuous frieze patterns*, illustrated in Figure 4.52. We can denote the top one as THRVG and the bottom one, which has an extra line below the band, as TV. Making continuous

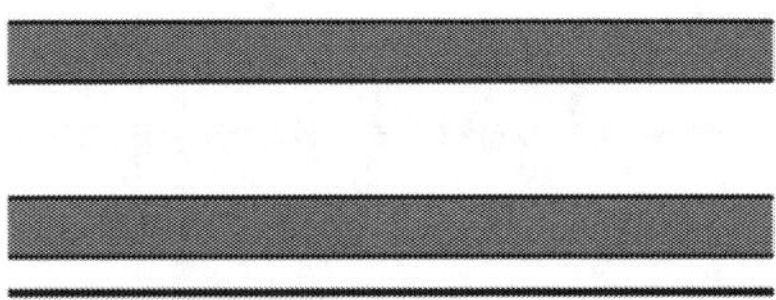

Figure 4.52. The two types of continuous frieze patterns.

horizontal translations turns any point of the design into a horizontal line. So a pattern consists of horizontal strips and lines. Thus every continuous frieze will have a vertical mirror at each point of the design. Adding in any of the other symmetries (H, R, or G) includes the others as well.

Problem 3.7 (c). (repeated) Find the five pairs of groups that can't be realized as two-color frieze patterns. Design frieze patterns using three or four colors for four of those types.

The five combinations of frieze patterns not among the two-color frieze patterns are TG/TG, THRVG/T, TRVG/T, THRVG/TG, and TRVG/TRVG. Figure 4.53 give examples of the first four using three or four colors. It might seem like we just need to keep drawing different patterns and perhaps adding colors until we find some number of colors compatible with the fifth combination. However, no matter how many colors we use, there is no frieze pattern of type TRVG/TRVG. A proof of an impossibility result like Theorem 4.8 is quite different from the other proofs we have considered. We need to understand more deeply what is possible in order to prove something is impossible. We will use a proof by contradiction; that is, we assume some such colored pattern exists and then show that would force some impossible situation. As a result, we conclude that the initial assumption of existence had to be incorrect.

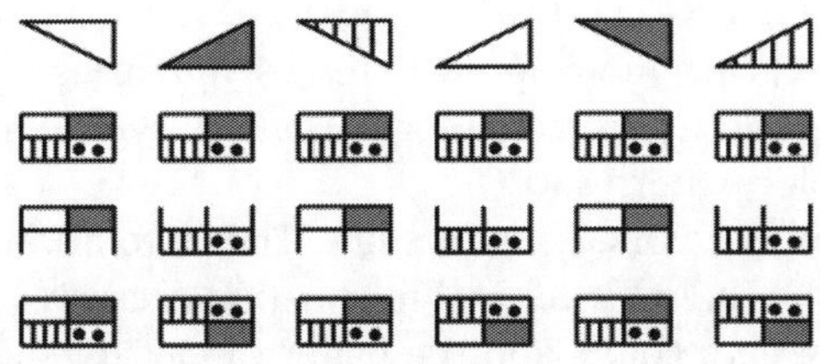

Figure 4.53. The three-color frieze pattern TG/TG and the four-color frieze patterns THRVG/T, TRVG/T, and THRVG/TG.

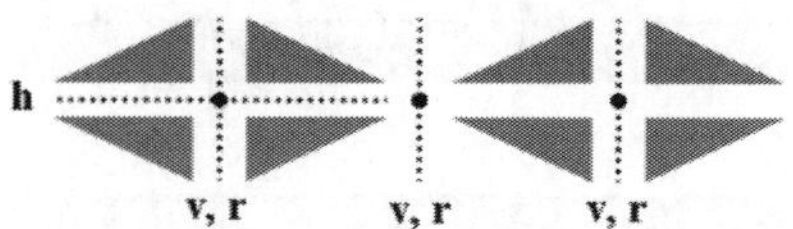

Figure 4.54. A THRVG pattern with rotations and reflections marked.

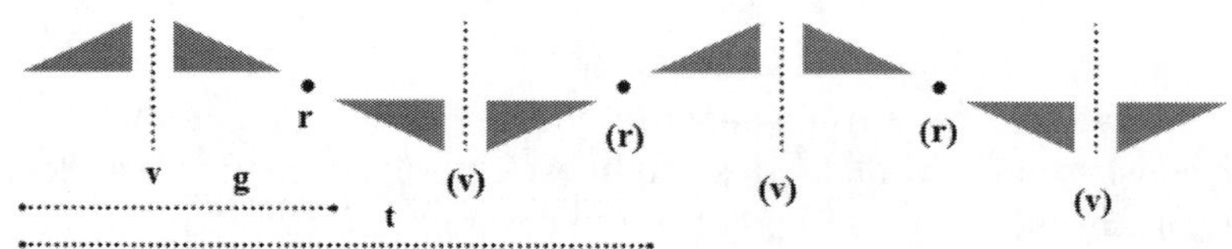

Figure 4.55. A TRVG pattern with some symmetries marked.

Theorem 4.8. No matter what n is ($n \geq 2$), there is no n-color frieze pattern TRVG/TRVG with a regular coloring.

Proof. Suppose for a contradiction that there is some pattern with some number of colors that has the classification TRVG/TRVG.

First let's analyze a TRVG pattern without worrying about colors. This pattern has no horizontal mirror reflection, so no center of rotation is on a line of vertical reflection. (Otherwise their composition would give a horizontal mirror, as in Figure 4.54.) Let v be a vertical mirror and r be a rotation that are as close together as any other such pair. As Figure 4.55 indicates, the length of the glide part of the shortest glide reflection (g) is twice the distance between v and r. Also, since there is no horizontal mirror, the length of the shortest translation (t) is twice the length of the shortest glide reflection. (For a THRVG pattern, as in Figure 4.54, the length of the shortest translation is equal to the length of the shortest glide.)

The pattern for TRVG is built using v and r. Now we consider cases with regard to color-preserving and color-switching symmetries.

Case 1. Both v and r preserve colors. Then all symmetries preserve colors and it is the one-color pattern TRVG.

Case 2. v preserves colors, but r does not. Then a rotation or other symmetry of the line of v is the line of a vertical mirror that preserves colors. That is, all vertical mirrors preserve colors and the pattern is TRVG/TV, regardless of how many colors there are.

Case 3. r preserves colors, but v does not. As in Case 2, all rotations preserve colors and the pattern is TRVG/TR, regardless of how many colors there are.

Case 4. Neither v nor r preserve colors. This is the only option that might give our TRVG/TRVG pattern. A center of rotation is halfway between two nearest lines of vertical reflection. Similarly, a line of vertical reflection is halfway between two nearest centers of rotation. (See Figure 4.55.) That means a rotation takes a vertical mirror that switches colors to another mirror that switches colors. Similarly, a vertical mirror takes a rotation that switched colors to a rotation that switches colors. So no vertical mirror and no rotation preserve colors. However, our assumption is that the pattern is TRVG/TRVG, which means there have to be color-preserving rotations and vertical mirrors. Contradiction. Hence the pattern TRVG/TRVG can't occur in any frieze pattern, regardless of the number of colors.   □

Remark. It is possible to give a shorter proof of Theorem 4.8 using abstract algebra, but it goes beyond the level of this book. While there is a proper subgroup of the TRVG group isomorphic (structurally identical) to the entire group, one can prove that such a subgroup can't be normal. And the color-preserving subgroup is always normal in the color group. See [30, 155–156].

Perhaps the most famous example of impossibility proofs arose from the ancient Greeks "questioning the answer" of some of their successes. They had, using only a straightedge and compass proven how to make many constructions. For instance, they knew how to bisect a general angle, construct a square with twice the area of a given square, and construct a square with the same area as a given polygon. Like mathematicians everywhere, once they solved those problems, they made up new variations. However, in spite of much effort, they and many later mathematicians were unsuccessful on three variations of these: *trisecting an angle, doubling a cube* (constructing a cube with twice the volume), and *squaring the circle* (constructing a square with the same area as a circle). It took over 2000 years and the deep insights of abstract algebra to prove the impossibility of these three constructions using these limited tools. For more on constructions, these problems, and the algebra needed to prove the impossibility, see [29, 12–17, 24–25] and [30, 277–284]. There is even a book [13] devoted to the too-many incorrect attempts at trisecting an angle, even after the impossibility was proven.

Problem 3.7 (d). (repeated) Look for examples of (some of) the seventeen different types of wallpaper patterns.

Figure 4.56 gives examples of each of the seventeen types of wallpaper patterns with the names of their groups of symmetries. There are 46 two-color wallpaper patterns, far too many to show here. (See [18, 406–413] for illustrations. See [29, 278–281] for aid in classifying these patterns using a flow chart.)

Artists from many cultures have delighted in making frieze and wallpaper patterns. Some anthropologists and archeologists, equipped with flow charts, have used the mathematical classification of these patterns the better to understand some cultures and their connections. Consider for instance, the two-color

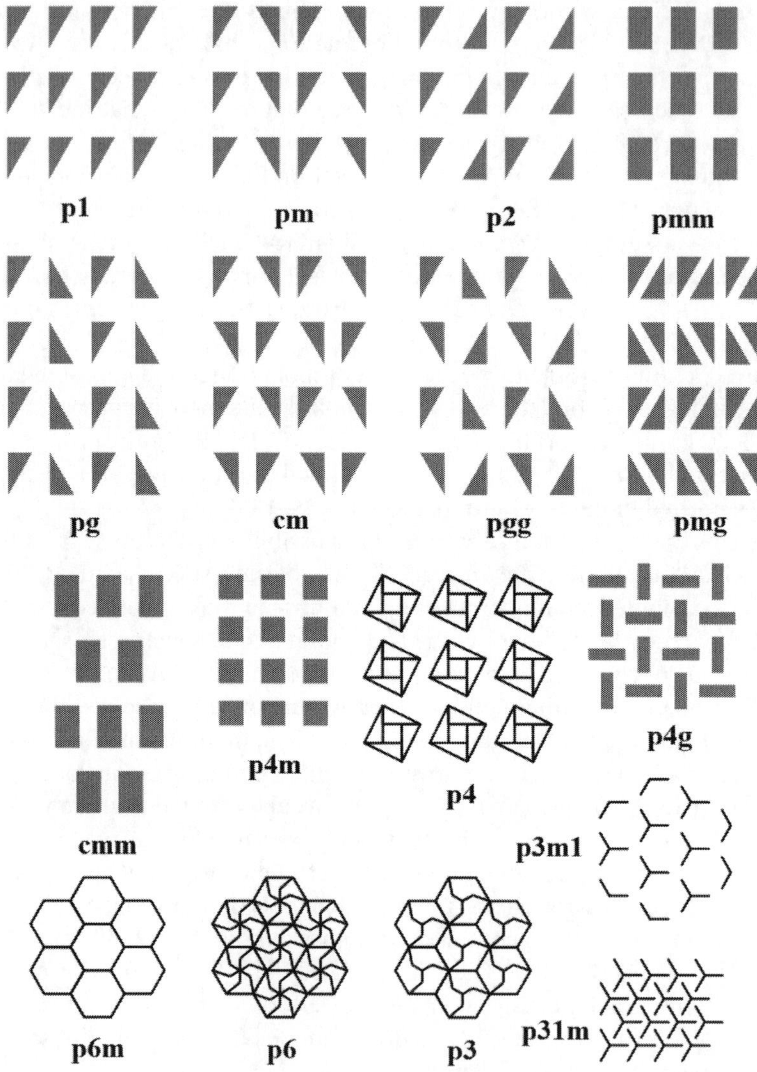

Figure 4.56. The seventeen types of wallpaper patterns.

wallpaper pattern from Egypt in Figure 4.57. This pattern mimics the threads of the warp and woof of a weaving pattern. If this type appeared on some ancient pottery, that could be evidence that the culture had developed weaving, even if no cloth survived as direct proof. The two-color weaving pattern type is p4g/cmm, as is Figure 4.57. See [9] for more information and examples from many cultures.

Figure 4.57. An Egyptian two-color wallpaper pattern.

Figure 4.58. One-color and two-color wallpaper patterns.

Exercise 4.7. Classify the wallpaper patterns in Figure 4.58. (The four patterns on the bottom are two-colored.)

The three-dimensional analogs of frieze patterns and wallpaper patterns are crystals. Real chemical crystals are, of course, finite but are modelled very well by infinite repeating patterns with translations in three directions. As noted in Section 3.5, Evgraf Fedorov classified all 230 groups of possible symmetries for mathematical crystals. He along with others were able to predict the atomic structure of crystals more than twenty years before X-ray crystallography was able to confirm this. Until 1984, all known chemical crystals fit into Fedorov's classification perfectly. But that year a team led by Dan Shechtman published evidence of a chemical compound whose crystallography behaved like that of a crystal but the crystal didn't have any translations. That compound and others like it are now called *quasicrystals*. In 2011, Shechtman won the Nobel Prize in Chemistry for this work. To the surprise of the chemists, mathematicians had already been "playing" with abstract versions of these. An early publication of the two-dimensional version, called a *Penrose tile*, appeared in a recreational mathematics journal in 1974. When the pattern of the quasicrystal was published in 1984, mathematicians could match it with a three-dimensional hyperplane cross section of a six-dimensional tessellation. (No one pretends to know what six dimensions have to do with real chemical quasicrystals.) Now mathematicians and chemists are collaborating on this active area of research. For more on Penrose tiles and quasicrystals, see [16] and [27].

Problem 3.7 (e). (repeated) Can every convex pentagon give a tessellation of the plane? If not, find some convex pentagons that will tessellate the plane. Repeat with hexagons.

A regular pentagon can't tessellate the plane because its external angles, all 108°, don't divide 360°. A regular hexagon with exterior angles of 120° will tessellate, giving the familiar honeycomb pattern. In 1918, K. Reinhardt (1895–1941) classified the three possible families of convex hexagons that can tile the plane. The possibilities for convex pentagons proved much more complicated with multiple mathematicians contributing. In 2017, Michäel Rao announced in an as-yet-unpublished paper that a computer assisted proof showed that there are fifteen families of convex pentagons. (See [1, 154–155].) Figure 4.59 gives several examples of convex pentagons that tile. These three are related to hexagons in different ways. See [1], [2, 43–48 and 75–76], [20], [21], and [26] for discussions of tessellations and the many pentagonal and hexagonal types.

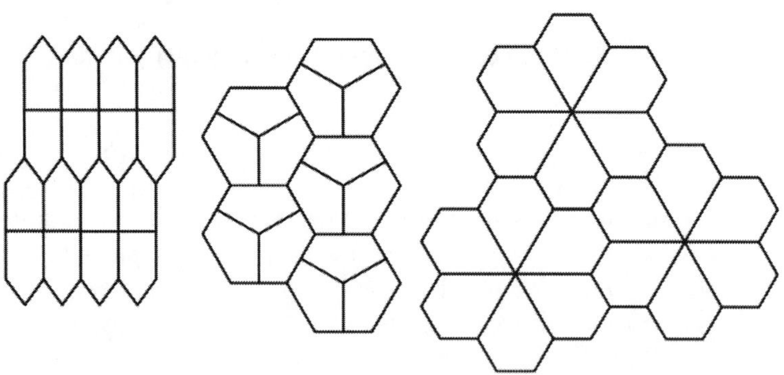

Figure 4.59. Some pentagonal tilings.

Problem 3.7 (f). (repeated) Find some regular colorings of the faces of the five regular polyhedra, shown in Figure 3.32, repeated here. Determine the size of the color-preserving subgroup and the color group for each regular coloring. The tetrahedron has four triangles, the cube has six squares, the octahedron has

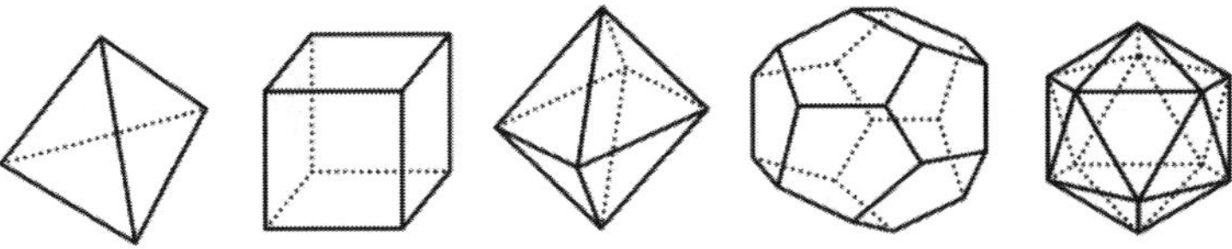

Figure 3.32. (repeated) The five regular polyhedra.

eight triangles, the dodecahedron has twelve pentagons, and the icosahedron has twenty triangles.

The boring cases, where every face is a different color or there is only one color, are always regular colorings of the regular polyhedra because we can map any face to any other face. When all the faces have different colors, the color-preserving subgroup has one element, the identity. Its color group is the same as the color-preserving and color groups when there is just one color. Its size depends on the regular polyhedron. For a tetrahedron, there are 24 symmetries; for the cube and octahedron, there are 48 symmetries; and for the dodecahedron and icosahedron, there are 120 symmetries. We will concentrate on colorings where the number of colors is between one and the number of faces. By Theorem 2.6, in a regular coloring, the size of the color group is a multiple of the number of faces and divides the total number of symmetries. The size of the color-preserving group divides the size of the color group.

Theorem 2.6 also forces the number of colors in a regular coloring to divide the number of faces. For the tetrahedron, the only option is to have two faces one color and the other two another color. In this case, the color-preserving group has four symmetries and the color group has eight. The other regular polyhedra are more interesting. The variety of regular colorings for most of them suggests some of the intricacies of the groups of symmetries of these polyhedra, even though their geometrical aspect may seem simple.

The cube with six faces has regular colorings with two or three colors. In a two-coloring the three faces that meet at the hidden vertex have one color (represented by dotted lines) and the three visible faces a different color (white) as in Figure 4.60 (a). The subgroup of color-preserving symmetries for this regular coloring has six symmetries and the color group has twelve symmetries. One three-coloring has opposite faces the same color, as in Figure 4.60 (b). It has eight color-preserving symmetries and forty-eight symmetries in its color group. Figure 4.60 (c) illustrates a different three-coloring in which each color covers two adjacent faces. The color-preserving subgroup has just one element, the identity, and the color group has six elements, all rotations.

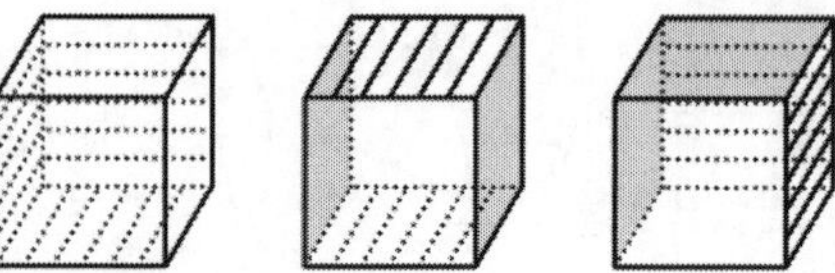

Figure 4.60. Regular colorings of a cube. (a) A two-coloring. (b) A three-coloring. (c) Another three-coloring.

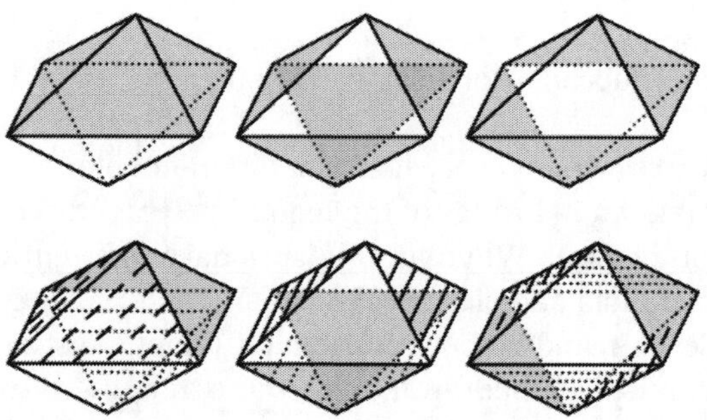

Figure 4.61. Regular two-colorings and four-colorings of a regular octahedron.

The octahedron with eight faces has a number of two-colorings and four-colorings. We show three two-colorings in the first row of Figure 4.61. For the one on the left, the four triangles around the top vertex are the same color and the four bottom triangles are another color. The color-preserving subgroup has eight symmetries and the color group has sixteen. A more interesting coloring is like a checkerboard as in the one in the middle: adjacent triangles are different colors. Its color-preserving group has twenty-four symmetries and its color group has forty-eight symmetries. The top right one has two adjacent triangles on the top of one color and the two triangles diametrically opposite them have the same color, while the other four faces are the other color. It has eight symmetries in its color-preserving subgroup and sixteen in its color group.

We can convert each of the two-colorings of an octahedron in the first row of Figure 4.61 into a four-coloring. For the leftmost two-coloring, recolor two adjacent triangles on the top with a third color and use a fourth color for the two triangles immediately below them. The color-preserving subgroup has just two symmetries and the color group has sixteen symmetries. Starting with the middle two-coloring, recolor two of the same colored triangles on the bottom

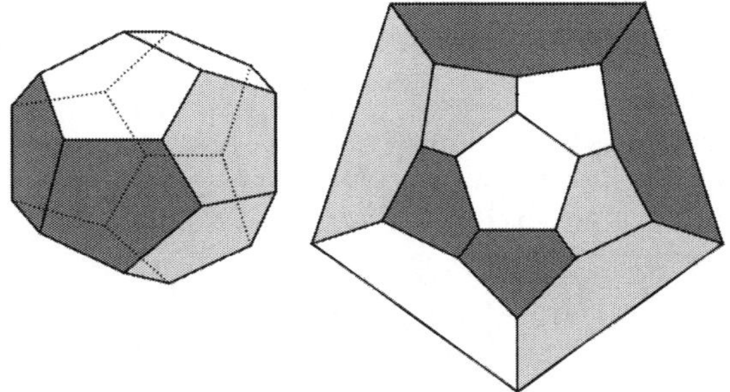

(a) A three-coloring of the dodecahedron.

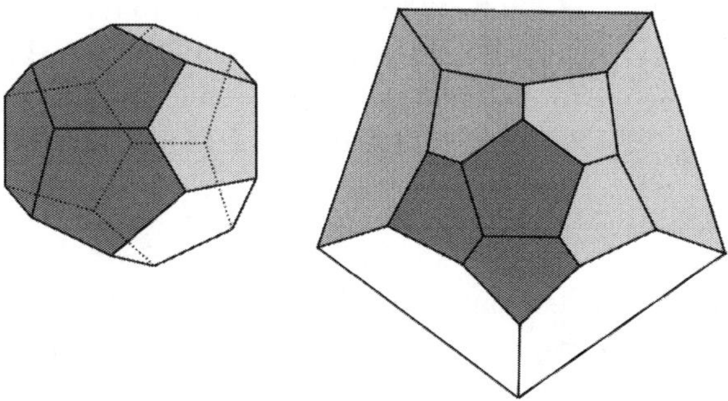

(b) A four-coloring of the dodecahedron.

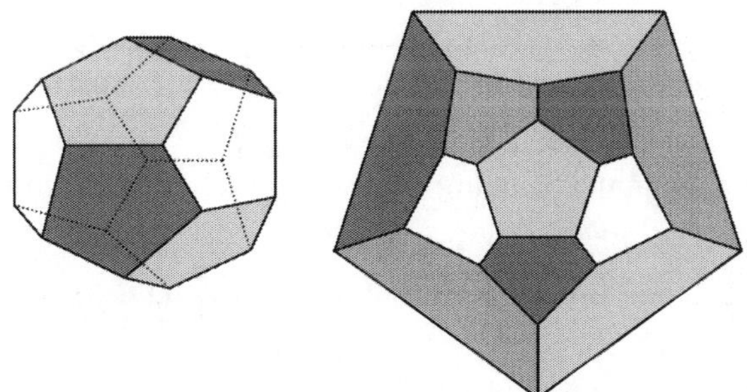

(c) Another four-coloring of the dodecahedron.

Figure 4.62.

with a third color and the other two triangles on the bottom with a fourth color. Its color-preserving subgroup has four symmetries and its color group has sixteen symmetries. The third four-coloring has only diametrically opposite faces the same color. Its color-preserving subgroup has only two symmetries, while its color group has forty-eight symmetries.

A regular dodecahedron with twelve faces has regular colorings with three, four, or six colors, but not two colors. To enable visualizing each of the colorings, we provide two representations of each polyhedron. The left one shows the colors on the visible faces. The other portrays a flatten net of all but the back face, which is colored white. In the three-coloring shown in Figure 4.62 (a), each color has two adjacent faces and the two faces opposite them. There are eight symmetries preserving the colors and twenty-four in the color group. The four-coloring of Figure 4.62 (b) has three faces sharing a vertex with the same color. Only the identity preserves all four colors, while the color group has twelve symmetries. Another more complicated four-coloring appears in Figure 4.62 (c). As with Figure 4.62 (b), only the identity preserves all four colors and there are twelve symmetries in the color group for Figure 4.62 (c). In the six-coloring (not shown), opposite faces have the same color. The color-preserving group for the six colors has two symmetries, and all 120 symmetries of the dodecahedron are in the color group.

A two-coloring can't be regular. We can easily color half the faces one color and the other six another color. However, to be a regular coloring, we need to have color-preserving symmetries taking any face to any other face of the same color. This means that each face has to have the same number of adjacent faces of the same color. Exercise 4.8 asks you to verify that none of the possibilities for a two-coloring are regular colorings.

Exercise 4.8. With a dodecahedron or the picture of one, explain why each of the following possible ways of two-coloring it fails to be a regular coloring.

(a) Each face has one adjacent face of the same color. Given one pair of faces the same color, how many faces are adjacent to them and so would have to be the same color?

(b) Each face has two adjacent faces of the same color. There are two options.

Case 1. There are two separated sets of three mutually adjacent faces of the same color. Use Figure 4.62 (b) to explain why this can't happen for both colors.

Case 2. The two faces the same color and adjacent to a given face are not adjacent to each other. Explain why the six faces of one color would need to form a ring to be a regular coloring. Can the other six faces also form a ring?

(c) Each face has three adjacent faces of the same color. Consider cases to explain why you can't do this with six faces of one color.

(d) Repeat part (c) with four or five adjacent faces of the same color.

There is a five-color regular coloring of the regular icosahedron with four faces of each color. The color group has 60 rotations and no reflections. (See Figure 4.63, which provides two representations.) The color-preserving group has only the identity in it. There is also a regular coloring with ten colors (not shown), with opposite faces the same color. The color-preserving group has one symmetry besides the identity, and the full color group has all 120 symmetries of the icosahedron. Analyses similar to Exercise 4.8 show that the icosahedron can't have a regular two-coloring or a regular four-coloring.

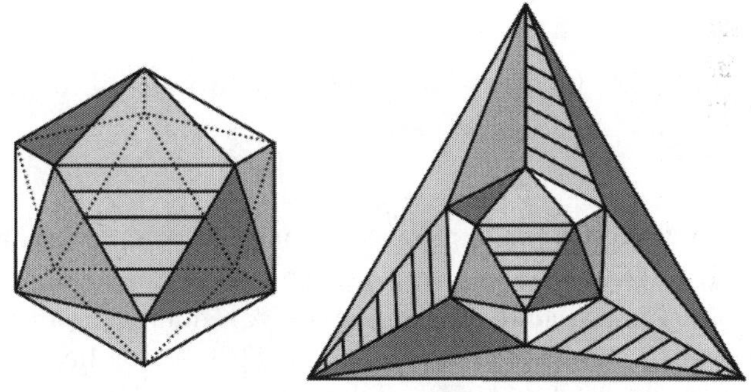

Figure 4.63. A regular five-coloring of a regular icosahedron.

4.6 Voronoi Diagrams

Problem 3.8 (a). (repeated) Explore generalizing Voronoi diagrams to three-dimensional Euclidean space or even higher dimensions. Will the regions still be convex?

Mathematicians, including Voronoi himself, have investigated Voronoi diagrams in higher dimensions. The following definition modifies the definition of Section 1.6 by replacing "the plane" with "n-dimensional Euclidean space."

Definition. Given sites $S_1, S_2, \ldots, S_n$ in n-dimensional Euclidean space, the *Voronoi region* of S_i is the set of all points X so that for all sites S_k with $k \neq i$, $d(S_i, X) < d(S_k, X)$, where $d(X, Y)$ is the (usual Euclidean) distance between the points X and Y. The Voronoi regions together with their boundaries form the *Voronoi diagram* of the sites.

In Section 3.6, given two sites A and B, the perpendicular bisector of the segment $\overline{AB}$ split the plane into two convex parts, the region closer to A and the region closer to B. In higher dimensions, the same thing happens. Figure 4.64

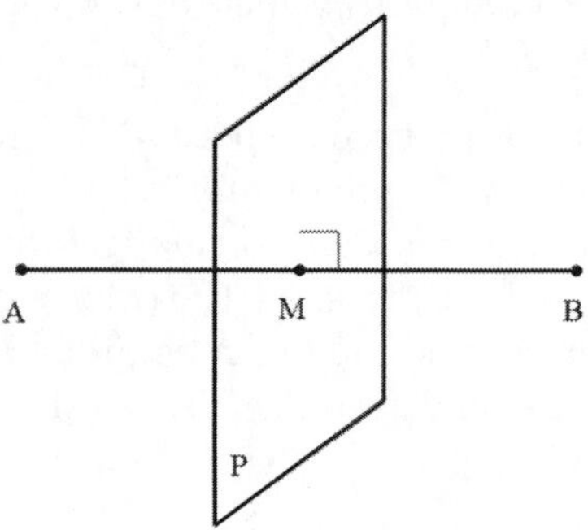

Figure 4.64. The plane P is perpendicular to $\overline{AB}$ through the midpoint M.

indicates the three-dimensional situation, where the perpendicular bisector is now a plane splitting the space into two convex half-spaces. Given multiple sites, we intersect these half-spaces to determine the Voronoi region for each site. The proof of Theorem 3.13 illustrates the elegance of abstract mathematics: it already applies to convex sets in any number of dimensions. So, Voronoi regions in any number of Euclidean dimensions are convex.

Figures 3.60 and 3.61 illustrated possible Voronoi regions in the plane when the sites form a lattice. In three dimensions, the regions will be polyhedra. When the sites in three dimensions form a lattice, these polyhedra can take interesting forms that tessellate space. Figure 4.65 illustrates one of the Voronoi regions, all

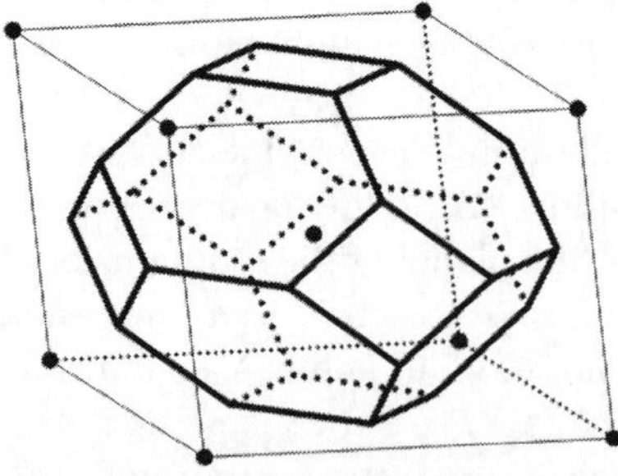

Figure 4.65. The Voronoi region for the site at the center of the shown cube, when the sites are vertices of a cubic lattice and the centers of those cubes.

truncated octahedra, where the sites are at the vertices of a cubic lattice along with the centers of each of those cubes. Fedorov found the five possible types of space filling polyhedra for such Voronoi regions in 1885. For more on this topic, including applications, see [23].

Problem 3.8 (b). (repeated) In Figures 3.65 and 3.66, determine the Voronoi boundary between C and D and the Voronoi boundary between E and F. Describe the possible shapes of Voronoi boundaries in the taxicab metric.

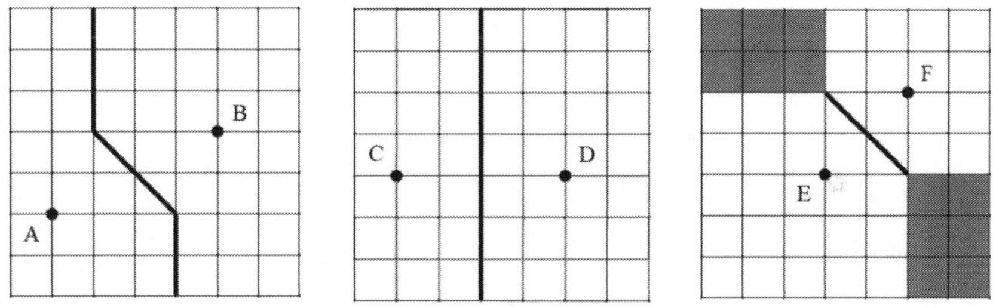

Figure 4.66. Voronoi boundaries in the taxicab metric. (a) The general setting. (b) The points have one coordinate in common. (c) The points are on a line with slope ± 1.

Figure 4.66 provides the boundaries to the figures in Exercise 3.5 and Problem 3.8 (b). It also represents the possible Voronoi boundaries in the taxicab metric. In the general case, the boundary is made of a line segment with slope ± 1 and two rays, either both horizontal or both vertical. In Figure 4.66 (a), the horizontal difference between points A and B is greater than the vertical difference. In case the differences are reversed, the rays will be horizontal. Figure 4.66 (b) represents the special cases when the two points have the same x-coordinate or y-coordinate. In these cases, the boundary is the Euclidean horizontal or vertical perpendicular bisector, respectively. When the line between the two points has slope ± 1, as in Figure 4.66 (c), the boundary includes a segment of the perpendicular bisector of the two points and two infinite quadrants of the plane.

Problem 3.8 (c). (repeated) Find the Voronoi boundaries in the taxicab metric if the sites are A $= (1, 1)$, B $= (3, 7)$, C $= (7, 5)$, D $= (5, 9)$, and E $= (10, 4)$. Compare with the Voronoi diagram for these points using the usual Euclidean distance.

The regions in the taxicab metric are not convex (at least in the Euclidean meaning of convex), whereas we know in Euclidean geometry the regions must be convex. The boundaries in the taxicab metric are made of segments and rays that are vertical, horizontal, or have slopes of ± 1. In Euclidean geometry, boundaries can have any slope (or be vertical). Perhaps the most glaring difference

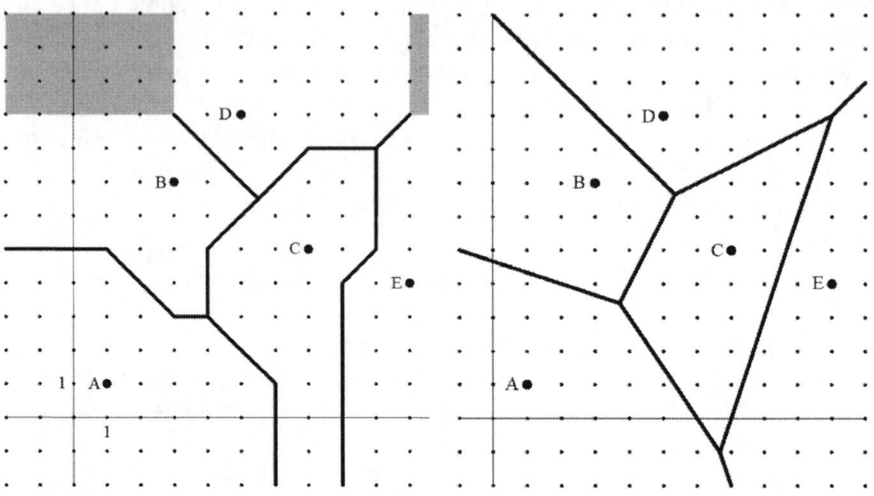

Figure 4.67. Voronoi boundaries for Problem 3.8 (c) in taxicab metric.

Figure 4.68. Voronoi diagram for Problem 3.8 (c) in Euclidean metric.

between Figures 4.67 and 4.68 is the shaded regions in Figure 4.67 indicating regions equidistant from B and D or from D and E. Also, it happens that none of the taxicab regions are bounded, whereas the Euclidean one has one bounded region. For more information on Voronoi diagrams, see [11, 98–117].

Elementary treatments of taxicab geometry consider lines to be as they are in Euclidean geometry. An alternative could be based on a modification of the common saying "a line segment is a shortest path between two points." (The more common misphrasing "a straight line is the shortest distance between two points" is inaccurate for two reasons. First, a line is infinitely long. More importantly, the distance between two points is a set number, so it can't be a path and it doesn't make much sense to talk about the shortest number.) If we define a line segment to be a shortest path, then the Voronoi regions in the taxicab metric are "taxicab convex" as in the proposed definition below: given any two points in a region, they are connected by some shortest path that stays within the region. Figure 4.69 indicates that some shortest path between two points in the taxicab metric only need to stay within the shaded rectangle determined by the points' coordinates. Shortest paths (in more formal terms *geodesics*) are important in differential geometry.

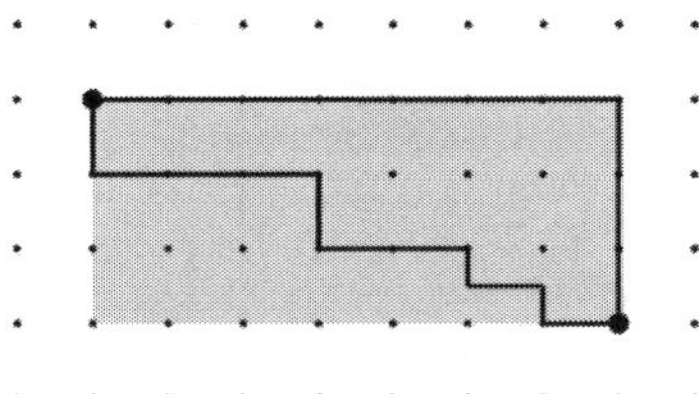

Figure 4.69. Two shortest paths in the taxicab metric.

Potential definition. A region is *taxicab convex* if for every two points of the region there is a shortest path between them that is a subset of the region.

This proposed definition of convex, however, doesn't satisfy some properties of convexity. In particular, an important theorem about convex sets (Theorem 3.13) is that their intersection is also convex. Figure 4.70 gives two taxicab convex regions whose shaded intersection is not even connected, let alone taxicab convex. While in principle we are free to define terms in mathematics, mathematicians need to consider the effects of their definitions. So, this potential definition is at best debatable.

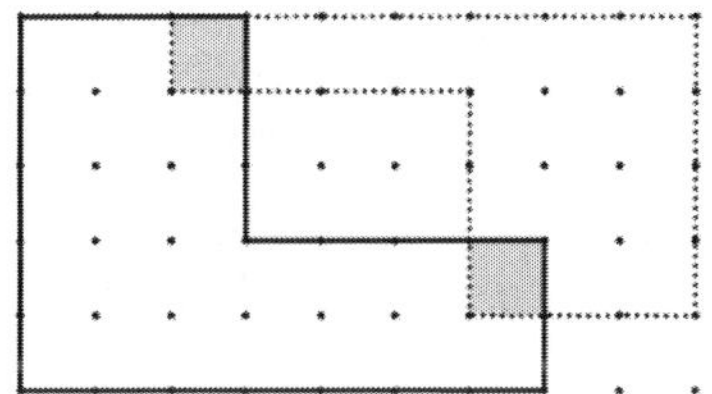

Figure 4.70. The intersection of "taxicab convex" regions can be "taxicab nonconvex."

It may seem anticlimactic to end this chapter with what seems to be a dead-end definition, but it is intentional. I hope the variety of variations of the original problems and the mathematics we have uncovered convinces you of how many ideas lie near at hand when we "question the answer." Perhaps you will develop a better definition of convex for the taxicab metric—or prove that no definition will have the properties we want for the idea of convex sets. I challenge you to push any of the topics of this book that interest you in new directions. Not all will succeed—that is the nature of explorations. Yet even dead-ends may redirect us in fruitful ways. I recall the many dead-ends I encountered in the research for my PhD dissertation. In retrospect without those failures, I wouldn't have developed the needed intuition and ability to ask new questions that ultimately led to success. The thrill of the adventure of developing something new is contagious. May you find that thrill.

Answers to Exercises

Exercise 1.1. (repeated) Draw other polygrams and compound polygons whose vertices are the vertices of a regular *n*-gon. Look for conditions determining whether you get a polygram or a compound polygon. What conditions do we need to set on the crossings so that the second drawing in Figure 1.9 is not a polygram?

Suppose that we make a circuit on the vertices of a regular polygon with *n* vertices by connecting every k^{th} vertex, as in Figure Ex.1. If *n* and *k* have a common divisor bigger than 1, then the circuit returns to the starting point before going to every vertex. Then we can repeat, starting from a missing vertex. This yields a compound polygon. In Figure Ex. 1, the first design has $n = 8$ and $k = 2$ or $k = 6$ and the fourth design has $n = 9$ and $k = 3$ or $k = 6$. But if the only common divisor of *n* and *k* is 1, then the circuit goes through each vertex and we have a polygram. In Figure Ex. 1, the second design has $n = 8$ and $k = 3$ or $k = 5$. The third design of Figure Ex. 1 has $n = 9$ and $k = 2$ or $k = 7$. The last design has $n = 9$ and $k = 4$ or $k = 5$. To ensure that we don't get other designs besides polygrams and compound polygons, we need to require that we skip over the same number of vertices each time in our circuits.

Figure Ex.1. Examples of polygrams and compound polygons.

Exercise 2.1. (repeated) To show that the candidates differ, we need to show that if $g_i \circ p_x = g_m \circ p_y$, then $i = m$ and $x = y$. Use colors to explain why we can't have $i \neq m$ when $g_i \circ p_x = g_m \circ p_y$. Now assume that $i = m$ and so $g_i \circ p_x = g_i \circ p_y$. Use the inverse g_i^{-1} to deduce $p_x = p_y$. Thus we have at least *nk* symmetries.

Since $g_i \circ p_x = g_m \circ p_y$, $g_i \circ p_x$ and $g_m \circ p_y$ must take c_1 to the same color. Also, p_x and p_y leave color c_1 as c_1 and g_i takes c_1 to c_i and g_m takes c_1 to c_m. Thus, $g_i = g_m$. This reduces the equality to $g_i \circ p_x = g_i \circ p_y$. Composing g_i^{-1} on the left of each side gives $g_i^{-1} \circ g_i \circ p_x = g_i^{-1} \circ g_i \circ p_y$. By the definition of an inverse, $g_i^{-1} \circ g_i$ is the identity ε and so $p_x = \varepsilon \circ p_x = \varepsilon \circ p_y = p_y$.

Exercise 2.2. (repeated) To show that there are no other symmetries, let h be any symmetry. It must take color c_1 to some color, say c_i. Now g_i also takes color c_1 to c_i. Show that $g_i^{-1} \circ h$ preserves color c_1 and so is one of the symmetries p_1, $p_2, \ldots, p_n$, say p_w. Show how to write h as a composition of g_i and that p_w.

We have $g_i^{-1} \circ h = p_w$. Composing g_i on the left of each side gives $g_i \circ g_i^{-1} \circ h = g_i \circ p_w$. As in Exercise 2.1, $g_i \circ g_i^{-1}$ is the identity, so $h = g_i \circ p_w$.

Exercise 3.1. (repeated) For a third-degree function, suppose $f(n) = an^3 + bn^2 + cn + d$. Simplify the difference $f(n) - f(n-1)$.

$$
\begin{aligned}
f(n) - f(n-1) &= an^3 + bn^2 + cn + d \\
&\quad - \left(a(n-1)^3 + b(n-1)^2 + c(n-1) + d \right) \\
&= an^3 + bn^2 + cn + d \\
&\quad - \left(a(n^3 - 3n^2 + 3n - 1) + b(n^2 - 2n + 1) + c(n-1) + d \right) \\
&= 3an^2 + (-3a + 2b)n + a - b + c.
\end{aligned}
$$

Exercise 3.2. (repeated) Use Exercise 3.1 to resolve Problem 2.1 (b) using a difference equation. That is, let $M_3(n)$ be the maximum number of three-dimensional regions determined by n planes. We saw that $M_3(n) = M_3(n-1) + M(n-1) = M_3(n-1) + (\frac{1}{2}(n-1)^2 + \frac{1}{2}(n-1) + 1)$. Deduce the formula we found for $M_3(n)$.

We have $M_3(n) - M_3(n-1) = (\frac{1}{2}(n-1)^2 + \frac{1}{2}(n-1) + 1) = \frac{1}{2}n^2 - \frac{1}{2}n + 1$. By Exercise 3.1, $\frac{1}{2}n^2 - \frac{1}{2}n + 1 = 3an^2 + (-3a + 2b)n + a - b + c$. From the n^2 terms, $a = \frac{1}{6}$. Then the n terms give $-\frac{1}{2} = -3(\frac{1}{6}) + 2b$, making $b = 0$. In turn the constant terms give $1 = \frac{1}{6} - 0 + c$, yielding $c = \frac{5}{6}$. Finally, we need to use a particular value of $M_3(n)$ to determine d. We know that $M_3(1) = 2$, so $2 = \frac{1}{6}(1)^3 - 0(1)^2 + \frac{5}{6}(1) + d$, and $d = 1$. We have same formula as we found in answering Problem 2.1 (b), namely $M_3(n) = \frac{1}{6}n^3 + \frac{5}{6}n + 1$.

Exercise 3.3. (repeated) Artists in many cultures have made frieze patterns from strands giving a layered effect, as in Figure 3.50 (repeated here). Classify the frieze pattern types of the patterns in Figure 3.50. Try your hand at designing other such patterns, including some with different symmetries.

The Turkish and Indian designs in Figure 3.50 have types TG and TR, respectively. These are the two most common frieze types for layered designs whose strands are linked together. All seven types are possible, but the designs are not always interesting. Figure Ex.2 has type THG, and Figure Ex.3 has type TRVG. If we replace the triangles of Figure Ex.3 with rectangles, we have type THRVG. If instead we omit the triangles that point down, the pattern becomes TV, and by having only every fourth triangle, the pattern is reduced to T.

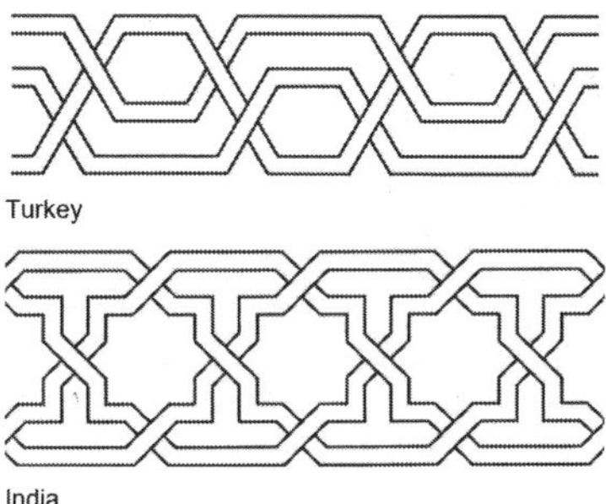

Figure 3.50. (repeated) Layered frieze patterns from Turkey and India.

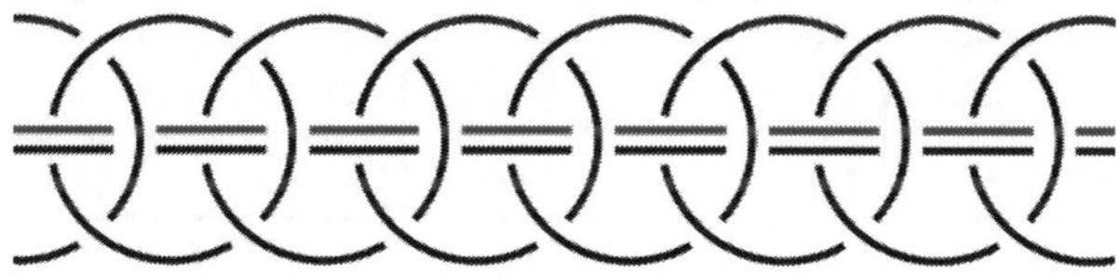

Figure Ex.2. A layered frieze pattern of type THG.

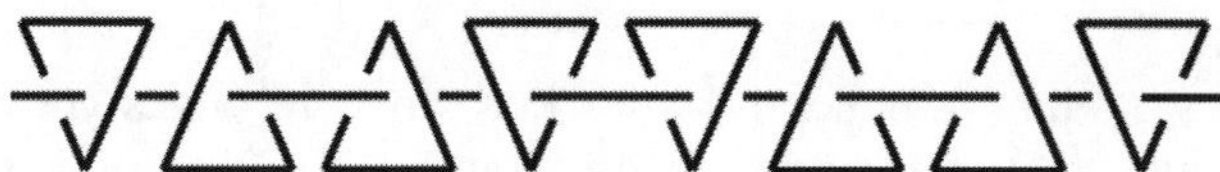

Figure Ex.3. A layered frieze pattern of type TRVG.

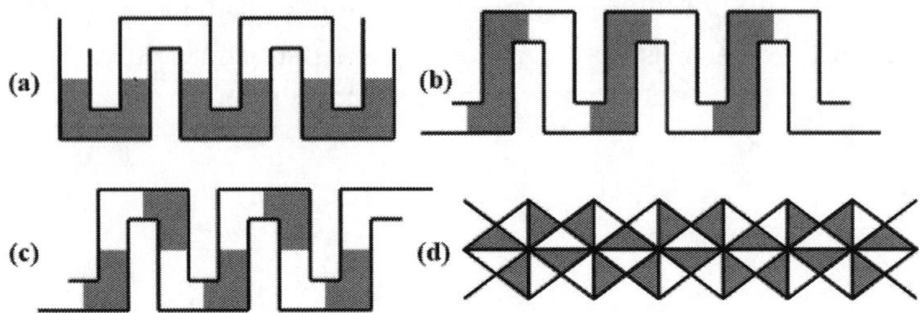

Figure 3.53. (repeated). Two-color frieze patterns. The patterns are (a) TRVG/TV, (b) TRVG/TR, (c) TRVG/TG, and (d) THRVG/TRVG.

Exercise 3.4. (repeated) Classify the two-color frieze patterns in Figure 3.53.

Exercise 3.5. (repeated) Find the Voronoi boundary between points A and B in Figure 3.62 using the taxicab metric.

(The answer appears in Figure 4.53 (a).)

Exercise 4.1. (repeated) Find and prove a formula for the volume of a *lattice box* whose eight vertices are lattice points in terms of the number of lattice points on its edges, surfaces, and interior as well as its vertices. (See Figure 4.13.)

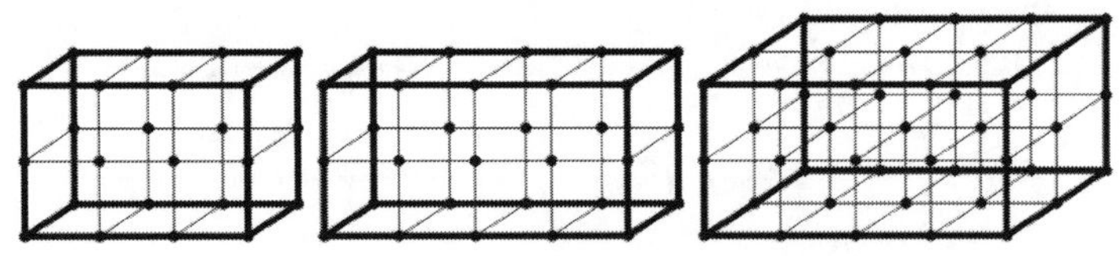

Figure 4.13. (repeated) Lattice boxes.

Proof. Consider a lattice box of dimensions $l \times w \times h$, which has volume lwh. Let V be the number of lattice points at vertices, E the number of lattice points on edges, S the number of lattice points on surfaces, and I the number of interior lattice points. The volume is $\frac{V}{8} + \frac{E}{4} + \frac{S}{2} + I$.

We always have $V = 8$, so $\frac{V}{8} = 1$. There are four edges with $l - 1$ lattice points, four with $w - 1$ lattice points, and four with $h - 1$ lattice points. Thus $E = 4(l + w + h - 3)$ and $\frac{E}{4} = l + w + h - 3$. There are two surfaces with

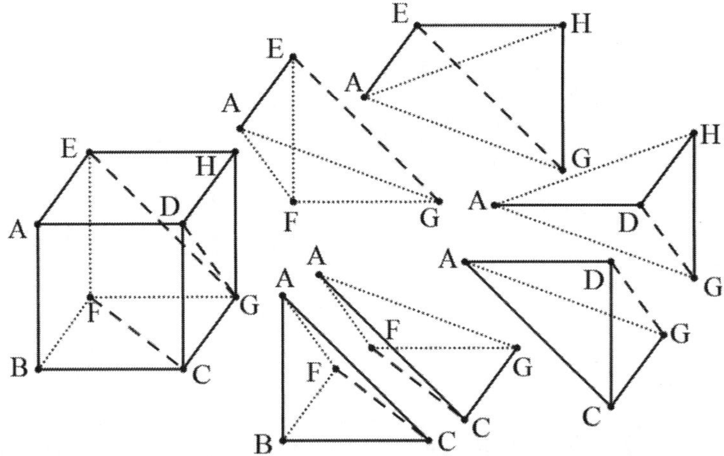

Figure 3.25. (repeated) Common vertex tetrahedralization of a cube.

$(l-1)(w-1)$ lattice points, two more with $(l-1)(h-1)$ lattice points, and two with $(w-1)(h-1)$ lattice points. Thus $S = 2(lw + lh + wh - 2l - 2w - 2h + 3)$ and $\frac{S}{2} = lw + lh + wh - 2l - 2w - 2h + 3$. There are $I = (l-1)(w-1)(h-1) = lwh - lw - lh - wh + l + w + h - 1$ interior lattice points. Then $\frac{V}{8} + \frac{E}{4} + \frac{S}{2} + I = lwh$, as required. $\qquad\square$

Exercise 4.2. (repeated) Find the area of regular tetrahedron whose vertices are four vertices of a unit cube, no two of which are adjacent to each other. Explain how this and the smallest lattice triangular pyramid give another way to show the impossibility of a three-dimensional Pick's theorem.

The unit cube has volume 1. We can "carve out" the regular tetrahedron by cutting off four corners of the cube, as in Figure 3.25. These corners each have a volume of $\frac{1}{3}bh = \frac{1}{3}(\frac{1}{2} \cdot 1 \cdot 1) \cdot 1 = \frac{1}{6}$. That leaves a volume of $\frac{1}{3}$ for the regular tetrahedron, which has no lattice points besides its four vertices. But one of the corners we cut off also has just four lattice points at its vertices with a different volume.

Exercise 4.3. (repeated) Look for other self-dual polyhedra besides pyramids. What properties do they have?

They need to have their number of vertices equal their number of faces. Also, when we match a face to a vertex, the number of vertices on a face has to match the number of faces sharing the vertex. One family, called elongated pyramids, amount to gluing an n-gonal pyramid on top of an n-gonal prism, as in Figure Ex.4 (a). It has n triangles, n rectangles, and one n-gon. There is another family, called diminished trapezohedra, with n triangles, n kites (quadrilaterals), and

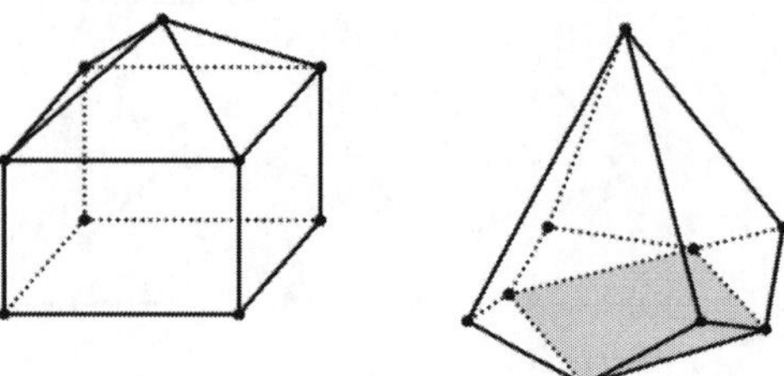

Figure Ex.4. (a) An elongated pyramid. (b) A diminished trapezohedron.

one *n*-gon. For these polyhedra the triangles are adjacent to the *n*-gon and the kites taper to the apex. (See Figure Ex. 4 (b).)

Exercise 4.4. (repeated) Explore whether the polyhedron with a hole in Figure 4.26 has a dual and, if so, whether their symmetries match and whether it is self-dual. Generalize.

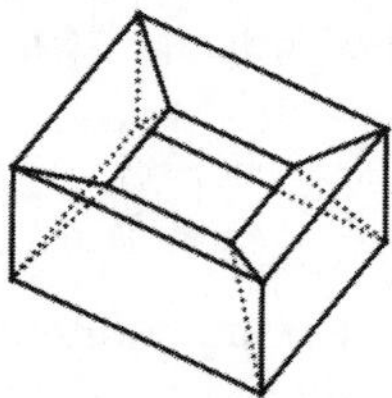

Figure 4.26. (repeated) A polyhedron with a hole.

Yes, the polyhedron in Figure 4.26 is self-dual. It has twelve faces, all quadrilaterals, and twelve vertices, all with four adjacent faces. There is a family of similar polyhedra with one hole. They have $n \times k$ faces, all quadrilaterals, and $n \times k$ vertices, each with four adjacent faces. The polyhedron in Figure 4.26 has $n = 4$ quadrilaterals around the hole and $k = 3$ in each band around the region where the air of an inner tube would be. The $n \times k$ polyhedron and its dual have the same group of $4n$ symmetries, matching the symmetries of an *n*-gonal prism.

Exercise 4.5. (repeated) Use Poincaré's theorem (Theorem 4.5) to verify that the Császár polyhedron with $V = 7$, $E = 21$, and $F = 14$ must have one hole. How can you determine that all of its faces must be triangles, that there are no diagonals, and that each vertex has six edges and so six triangles adjacent to it?

From Poincaré's theorem, $V - E + F = 2 - 2H$, we have $7 - 21 + 14 = 2 - 2H$, giving $H = 1$. Every face has to have at least three edges, and each edge is on two faces. Then we have $3F \leq 2E$, with equality just when every face is a triangle. But $3(7) = 2(14)$, so each face is a triangle. With seven vertices, there are $\binom{7}{2} = \frac{7 \cdot 6}{2} = 21$ segments connecting them. There are that many edges, so there are no diagonals and each vertex has to have six edges, one to each other vertex. A vertex has as many faces adjacent to it as edges. So, each vertex has six faces adjacent to it.

Exercise 4.6. (repeated) Use the reasoning in the previous paragraph to conjecture the maximum number of points in four-dimensional taxicab geometry for five and six distances.

For five distances, there are 456 points. We use points with coordinates whose absolute values add to 1, 3, or 5. We already saw that there are 96 points for the absolute values adding to 1 or 3. Here is a list of ways to get a sum of 5:

± 5 with three 0s (8 ways)
± 4, ± 1, and two 0s (48 ways)
± 3, ± 2, and two 0s (48 ways)
± 3, ± 1 twice, and 0 (96 ways)
± 2 twice, ± 1, and 0 (96 ways)
± 2, ± 1 three times (64 ways)

For six distances, there are 833 points. We use points with coordinates whose absolute values add to 0, 2, 4, or 6. We already saw that there are 225 points for the absolute values adding to 0, 2, or 4. Here is a list of ways to get a sum of 6:

± 6 with three 0s (8 ways)
± 5, ± 1, and two 0s (48 ways)
± 4, ± 2, and two 0s (48 ways)
± 4, ± 1 twice, and 0 (96 ways)
± 3 twice and two 0s (24 ways)
± 3, ± 2, ± 1, and 0 (192 ways)
± 3, ± 1 three times (64 ways)
± 2 three times and 0 (32 ways)
± 2 twice, ± 1 twice (96 ways)

Exercise 4.7. (repeated) Classify the wallpaper patterns in Figure 4.58. (The four patterns on the bottom are two-colored.)

The patterns are p4g, p6, pmg, p4m, p6/p3, p4/p4, pmg/pm, and pgg/pg.

Figure 4.58. (repeated) One-color and two-color wallpaper patterns.

Exercise 4.8. (repeated) With a dodecahedron or the picture of one, explain why each of the following possible ways of two-coloring it fails to be a regular coloring.

(a) Each face has one adjacent face of the same color. Given one pair of faces the same color, how many faces are adjacent to them and so would have to be the same color?

The two adjacent faces of one color have six faces adjacent to them, which would have to be the other color. But then each of these faces would have two adjacent faces of the same color. Then we couldn't interchange colors.

(b) Each face has two adjacent faces of the same color. There are two options.

Case 1. There are two separated sets of three mutually adjacent faces of the same color. Use the middle dodecahedron of Figure 4.62 to explain why this can't happen for both colors.

It is possible to have four sets of three mutually adjacent faces, but as in the middle dodecahedron of Figure 4.62 (b), each set is adjacent to all of the others. Then some faces would have more than two adjacent faces of the same color.

Case 2. The two faces adjacent to a given face are not adjacent. Explain why the six faces of one color would need to form a ring to be a regular coloring. Can the other six faces form a ring?

Suppose face 2 is adjacent to faces 1 and 3, 3 adjacent to 2 and 4, and so on. Then face 6 would have to be adjacent to faces 5 and 1, making a ring. The ring of faces of one color would separate the remaining six faces into two regions, one region on each side of the ring.

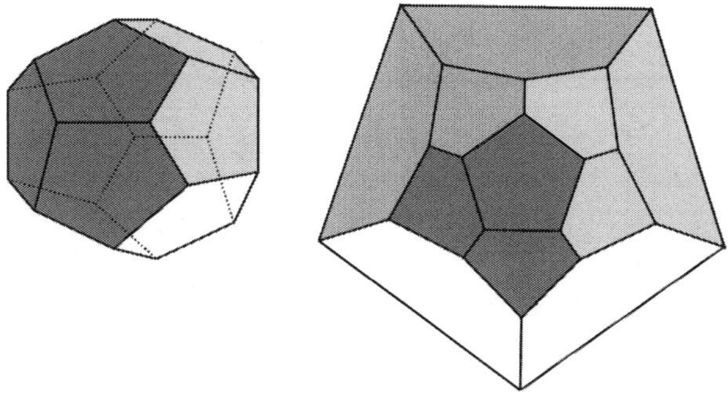

Figure 4.62 (b). (repeated) A four-coloring of the dodecahedron.

(c) Each face has three adjacent faces of the same color. Consider cases to explain why you can't do this with six faces of one color.

Figure Ex.5 shows the two ways for the darker face to have three lighter adjacent faces. On the left, two of these lighter faces are adjacent to each other but not to the third. On the right, one of these lighter faces is adjacent to both of the others. Either way, each of these three lighter faces needs to have three adjacent faces also of the same color and then these new faces will need three adjacent faces of the same color. We end up with more than six faces of the same color.

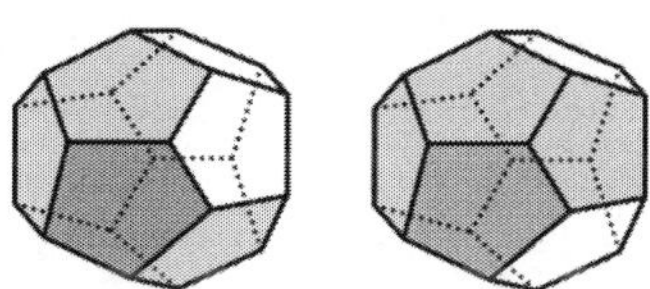

Figure Ex.5. Two ways for a face to have three adjacent faces.

(d) Repeat part (c) with four or five adjacent faces of the same color.

The same reasoning applies here as in part (c), forcing more than six faces of the same color.

Suggested Reading

C. Adams, *The tiling book: An introduction to the mathematical theory of tilings*, American Mathematical Society, Providence, RI, 2022.

The wonderful illustrations of this book will entice any reader to explore the variety of tilings. The writing is clear, but defines and uses some more advanced mathematical concepts than studied in high school.

C. Alsina and R. B. Nelson, *A panoply of polygons*, The Dolciani Mathematical Expositions, vol. 58, American Mathematical Society, Providence, RI, 2023.

An engaging read for general readers about all things polygonal.

D. Barnette, *Map coloring, polyhedra, and the four-color problem*, The Dolciani Mathematical Expositions, vol. 8, Mathematical Association of America, Washington, DC, 1983.

This readable book introduces the ideas leading to the proof of the four-color theorem.

A. Beck, M. N. Bleicher, and D. W. Crowe, *Excursions into mathematics*, The millennium edition, with a foreword by Martin Gardner, A. K. Peters, Ltd., Natick, MA, 2000.

In addition to many other interesting topics, this accessible book has an excellent exploration of many aspects of Euler's formula, map coloring, and polyhedra.

D. Crowe and D. Washburn, *Symmetries of culture: Theory and practice of plane pattern analysis*, University of Washington Press, Seattle, Washington, 1988.

This mathematician and anthropologist pioneered the use of symmetry analysis in anthropology and archeology. This book provides an anthropological perspective, detailed explanation of the mathematics for nonmathematicians, and hundreds of examples from cultures around the world.

K. Devlin, *The millennium problems: The seven greatest unsolved mathematical puzzles of our time*, Basic Books, New York, 2002.

In 2000, the Clay Institute, in honor of the new millennium, offered $1,000,000 prizes for published solutions of what they considered the most important unsolved mathematics conjectures. This well written book describes these deep open questions in accessible language.

L. Fortnow, *The golden ticket: P, NP and the search for the impossible*, Princeton University Press, Princeton, NJ, 2013.

Chapter 3 of this book explains the ideas of P and NP. Chapter 4 discusses NP-complete. It uses readily understood language and examples.

M. Gardiner, *Penrose tiles to trapdoor ciphers*, Mathematical Association of America, Washington, DC, 1997.

Reprints and updating of some of Martin Gardiner's famous, accessible, and engaging columns from *Scientific American*. The first one on Penrose tiles was written in 1977, a few years after Penrose first published his work. The second one on Penrose tiles came ten years later after much mathematical work and the then-recent discovery of quasicrystals.

B. Grünbaum and G. C. Shephard, *Tilings and patterns*, W. H. Freeman and Company, New York, 1987.

While this book is intended for professionals, it has thousands of intriguing illustrations that can stimulate anyone interested in geometric patterns.

P. Hoffman, *The man who loved only numbers: The story of Paul Erdős and the search for mathematical truth*, Hyperion Books, New York, 1998.

This delightful biography recounts many of the stories of this eccentric mathematician, as well as his important contributions to mathematics.

K. Miyazaki, *An adventure in multidimensional space: The art and geometry of polygons, polyhedra, and polytopes* (Henry Crapo, ed.), translated by Miyazaki, with forewords by Buckminster Fuller and H. S. M. Coxeter; With an afterword by Michio Maekawa, A Wiley-Interscience Publication, John Wiley & Sons, Inc., New York, 1986.

You can find pictures and more about these shapes, in nature, in computer graphics, and in other media.

D. Schattschneider, Tiling the plane with congruent pentagons, Math. Mag. **51** (1978), no. 1, 29–44.

This engaging article describes some of the history as well as the known families of examples and the mathematics of tilings.

T. M. Thompson, *From error-correcting codes through sphere packings to simple groups*, Carus Mathematical Monographs, vol. 21, Mathematical Association of America, Washington, DC, 1983.

This accessible book reveals some of the deep connections of these seemingly unrelated topics.

M. J. Wenninger, *Polyhedron models*, Cambridge University Press, London-New York, 1971.

If you want to build three-dimensional models, this is the book to get you started. In addition to pictures, you will find instructions and patterns for making the models.

R. Wilson, *Four colors suffice: How the map problem was solved*, Princeton Science Library, Princeton University Press, NJ, 2014.

This very readable book tells the engaging story of the mathematics and the mathematicians leading to the proof of the four-color map theorem. It also enjoys connecting other mathematical ideas to this involved story.

Bibliography

[1] C. Adams, *The tiling book: An introduction to the mathematical theory of tilings*, American Mathematical Society, Providence, RI, [2022] ©2022. MR4567741

[2] C. Alsina and R. B. Nelsen, *A panoply of polygons*, The Dolciani Mathematical Expositions, vol. 58, American Mathematical Society, Providence, RI, [2023] ©2023. MR4548176

[3] V. Balaji, O. Edwards, A. M. Loftin, S. Mcharo, L. Phillips, A. Rice, and B. Tsegaye, *Sets in $\mathbb{R}^d$ determining k taxicab distances*, Involve **13** (2020), no. 3, 487–509, DOI 10.2140/involve.2020.13.487. MR4129396

[4] W. W. R. Ball, *Mathematical Recreations and Essays*, Revised by H. S. M. Coxeter, The Macmillan Company, New York, 1947. MR19629

[5] D. Barnette, *Map coloring, polyhedra, and the four-color problem*, The Dolciani Mathematical Expositions, vol. 8, Mathematical Association of America, Washington, DC, 1983. MR741465

[6] A. Beck, M. N. Bleicher, and D. W. Crowe, *Excursions into mathematics*, The millennium edition, With a foreword by Martin Gardner, A K Peters, Ltd., Natick, MA, 2000. MR1744676

[7] R. Courant and H. Robbins, *What is mathematics?: An elementary approach to ideas and methods*, Oxford University Press, New York, 1979. MR552669

[8] H. S. M. Coxeter, *Introduction to geometry*, 2nd ed., John Wiley & Sons, Inc., New York-London-Sydney, 1969. MR346644

[9] D. Crowe and D. Washburn, *Symmetries of culture: Theory and practice of plane pattern analysis*, University of Washington Press, Seattle, Washington, 1988.

[10] Á. Császár, *A polyhedron without diagonals*, Acta Univ. Szeged. Sect. Sci. Math. **13** (1949), 140–142. MR35029

[11] S. L. Devadoss and J. O'Rourke, *Discrete and computational geometry*, Princeton University Press, Princeton, NJ, 2011. MR2790764

[12] K. Devlin, *The millennium problems: The seven greatest unsolved mathematical puzzles of our time*, Basic Books, New York, 2002. MR1930195

[13] U. Dudley, *A budget of trisections*, Springer-Verlag, New York, 1987, DOI 10.1007/978-1-4419-8538-5. MR913937

[14] Euclid, *The Thirteen Books of the Elements*, Dover, New York, 1956.

[15] L. Fortnow, *The golden ticket: P, NP, and the search for the impossible*, Princeton University Press, Princeton, NJ, 2013, DOI 10.1515/9781400846610. MR3026063

[16] M. Gardiner, *Penrose tiles to trapdoor ciphers*, Mathematical Association of America, Washington, DC, 1997.

[17] J. E. Goodman and J. O'Rourke (eds.), *Handbook of discrete and computational geometry*, CRC Press Series on Discrete Mathematics and its Applications, CRC Press, Boca Raton, FL, 1997. MR1730156

[18] B. Grünbaum and G. C. Shephard, *Tilings and patterns*, W. H. Freeman and Company, New York, 1987. MR857454

[19] P. Hoffman, *The man who loved only numbers: The story of Paul Erdős and the search for mathematical truth*, Hyperion Books, New York, 1998. MR1666054

[20] R. B. Kershner, *On paving the plane*, Amer. Math. Monthly **75** (1968), 839–844, DOI 10.2307/2314332. MR236822

[21] C. Mann, J. McLoud-Mann, and D. Von Derau, *Convex pentagons that admit i-block transitive tilings*, Geom. Dedicata **194** (2018), 141–167, DOI 10.1007/s10711-017-0270-9. MR3798065

[22] K. Miyazaki, *An adventure in multidimensional space: The art and geometry of polygons, polyhedra, and polytopes* (Henry Crapo, ed.), translated by K. Miyazaki, With forewords by Buckminster Fuller and H. S. M. Coxeter; With an afterword by Michio Maekawa, A Wiley-Interscience Publication, John Wiley & Sons, Inc., New York, 1986. MR874530

[23] A. Okabe, *Spatial Tessellations : Concepts and Applications of Voronoi Diagrams, 2*nd ed., John Wiley & Sons, Inc., New York, 2000.

[24] J. O'Rourke, *Art gallery theorems and algorithms*, International Series of Monographs on Computer Science, The Clarendon Press, Oxford University Press, New York, 1987. MR921437

[25] J. E. Reeve, *On the volume of lattice polyhedra*, Proc. London Math. Soc. (3) **7** (1957), 378–395, DOI 10.1112/plms/s3-7.1.378. MR95452

[26] D. Schattschneider, *Tiling the plane with congruent pentagons*, Math. Mag. **51** (1978), no. 1, 29–44, DOI 10.2307/2689644. MR493766

[27] M. Senechal, *Quasicrystals and geometry*, Cambridge University Press, Cambridge, 1995. MR1340198

[28] T. Q. Sibley, *Foundations of Mathematics*, John Wiley & Sons, Inc., Hoboken, NJ, 2009.

[29] T. Q. Sibley, *Thinking geometrically: A survey of geometries*, MAA Textbooks, Mathematical Association of America, Washington, DC, 2015. MR3362805

[30] T. Q. Sibley, *Thinking algebraically—an introduction to abstract algebra*, AMS/MAA Textbooks, vol. 65, MAA Press, Providence, RI; American Mathematical Society, Providence, RI, [2021] ©2021, DOI 10.1090/text/065. MR4389378

[31] R. P. Stanley, *Enumerative combinatorics. Vol. 2*, With a foreword by Gian-Carlo Rota and appendix 1 by Sergey Fomin, Cambridge Studies in Advanced Mathematics, vol. 62, Cambridge University Press, Cambridge, 1999, DOI 10.1017/CBO9780511609589. MR1676282

[32] L. Stoekmeyer, *Planar 3-colorability is polynomial complete*, SIGACT News (1973), 19–25.

[33] T. M. Thompson, *From error-correcting codes through sphere packings to simple groups*, Carus Mathematical Monographs, vol. 21, Mathematical Association of America, Washington, DC, 1983. MR749038

[34] M. J. Wenninger, *Polyhedron models*, Cambridge University Press, London-New York, 1971. MR467493

[35] M. J. Wenninger, *Dual models*, With a foreword by John Skilling, Cambridge University Press, Cambridge, 1983, DOI 10.1017/CBO9780511569371. MR730208

[36] R. Wilson, *Four colors suffice: How the map problem was solved*, Revised color edition of the 2002 original, with a new foreword by Ian Stewart, Princeton Science Library, Princeton University Press, Princeton, NJ, 2014. MR3235839

Index

Terms

People